Lecture Notes in Artificial Intelligence   11948

Subseries of Lecture Notes in Computer Science

Series Editors

Randy Goebel
University of Alberta, Edmonton, Canada
Yuzuru Tanaka
Hokkaido University, Sapporo, Japan
Wolfgang Wahlster
DFKI and Saarland University, Saarbrücken, Germany

Founding Editor

Jörg Siekmann
DFKI and Saarland University, Saarbrücken, Germany

Michelangelo Ceci · Corrado Loglisci ·
Giuseppe Manco · Elio Masciari ·
Zbigniew Ras (Eds.)

New Frontiers
in Mining Complex Patterns

8th International Workshop, NFMCP 2019
Held in Conjunction with ECML-PKDD 2019
Würzburg, Germany, September 16, 2019
Revised Selected Papers

 Springer

Editors
Michelangelo Ceci ⓘ
University of Bari Aldo Moro
Bari, Italy

Giuseppe Manco ⓘ
CNR-ICAR
Rende, Italy

Zbigniew Ras ⓘ
University of North Carolina
Charlotte, NC, USA

Corrado Loglisci ⓘ
University of Bari Aldo Moro
Bari, Italy

Elio Masciari ⓘ
Federico II University
Naples, Italy

ISSN 0302-9743   ISSN 1611-3349 (electronic)
Lecture Notes in Artificial Intelligence
ISBN 978-3-030-48860-4   ISBN 978-3-030-48861-1 (eBook)
https://doi.org/10.1007/978-3-030-48861-1

LNCS Sublibrary: SL7 – Artificial Intelligence

This Springer imprint is published by the registered company Springer Nature Switzerland AG
The registered company address is: Gewerbestrasse 11, 6330 Cham, Switzerland

Preface

Modern automatic systems are able to collect huge volumes of data, often with a complex structure (e.g. multi-table data, network data, web data, time series and sequences, trees and hierarchies). Massive and complex data pose new challenges for current research in Data Mining. Specifically, they require new models and methods for their storage, management, and analysis, in order to deal with the following complexity factors:

– Data with a complex structure (e.g. multi-relational, time series and sequences, networks, and trees) as input or output of the data mining process
– Data collections with many examples and/or many dimensions, where data may be processed in (near) real time
– Partially labeled data
– Data which arrive continuously as a stream, at high rate, subject to concept drift

The 8th International Workshop on New Frontiers in Mining Complex Patterns (NFMCP 2019) was held in Würzburg, Germany in conjunction with the European Conference on Machine Learning and Principles and Practice of Knowledge Discovery in Databases (ECML-PKDD 2019) on September 16, 2019. The purpose of this workshop was to bring together researchers and practitioners of Data Mining who are interested in the latest developments in the analysis of complex and massive data sources, such as blogs, event or log data, medical data, spatio-temporal data, social networks, mobility data, sensor data, and streams. The workshop was aimed at discussing and introducing new algorithmic foundations and representation formalisms in complex pattern discovery. Finally, it encouraged the integration of recent results from existing fields, such as Statistics, Machine Learning, and Big Data Analytics. This book features a collection of revised and significantly extended versions of papers accepted for presentation at the workshop. These papers went through a rigorous review process to ensure compliance with Springer's high-quality publication standards. The individual contributions of this book illustrate advanced Data Mining techniques that take advantage of the informative richness of both complex data and massive data for efficient and effective identification of complex information units present in such data.

The book is composed of nine chapters.

Chapter 1 proposes a framework, consisting of generic transformations, that allows for the combination of state-of-the-art time series representation, pattern mining, and pattern-based anomaly detection algorithms.

Chapter 2 deals with the problem of privacy preserving frequent pattern mining and proposes a heuristic approach for sensitive pattern hiding based on the selective deletion of items.

Chapter 3 proposes a novel survival analysis modeling approach based on gradient boosting using bagged trees as base learners. The proposed approach is shown to have higher predictive power while maintaining full interpretability.

Chapter 4 introduces a generalized neural network-based recommender framework that allows for the inclusion of more elaborate information from various data sources.

Chapter 5 deals with the problem of guaranteeing the interpretability in Graph Convolutional Neural Networks and proposes an approach for estimating the discriminative power of graph nodes from the models learned by a deep graph convolutional method.

Chapter 6 studies the problems that may arise in interleaved test-then-train evaluations when detecting concept drifts in data streams. It proposes an approach combining weighted soft voting and unsupervised drift detection to reduce the dependency on labels during model construction.

Chapter 7 focuses on Target-based Sentiment Analysis, i.e., the problem of identifying target-specific aspect words and opinion words within textual data.

Chapter 8 presents a systematic literature review of recent research dealing with customer purchase prediction in the E-commerce context. The authors propose a novel analytical framework and a research agenda in the field.

Finally, chapter 9 addresses the extraction of line parameters from spectrograms for audio data, recorded via cars passing by an audio recorder. The goal is to use these parameters to detect the speed behavior of the vehicles.

We would like to thank all the authors who submitted papers for publication in this book and all the workshop participants and speakers. We are also grateful to the members of the Program Committee and external referees for their excellent work in reviewing submitted and revised contributions with expertise and patience. We would like to thank Giorgiana Ifrim for her invited talk on "Effective Linear Models for Learning with Sequences and Time Series". A special thanks is due to both the ECML-PKDD workshop chairs and to the ECML-PKDD organizers who made this event possible. Last but not least, we thank Alfred Hofmann and Aliaksandr Birukou from Springer for his continuous support.

April 2020

Michelangelo Ceci
Corrado Loglisci
Giuseppe Manco
Elio Masciari
Zbigniew Ras

Organization

Program Chairs

Michelangelo Ceci	University of Bari Aldo Moro, Italy
Corrado Loglisci	University of Bari Aldo Moro, Italy
Giuseppe Manco	ICAR-CNR, Italy
Elio Masciari	Federico II University of Naples, Italy
Zbigniew Ras	University of North Carolina at Charlotte, USA

Program Committee

Petr Berka	University of Economics of Prague, Czech Republic
Jorge Bernardino	ISEC, Polytechnic Institute of Coimbra, Portugal
Carmela Comito	CNR-ICAR, Italy
Roberto Corizzo	American University, USA
Hadi Fanaee Tork	University of Oslo, Norway
Bettina Fazzinga	CNR-ICAR, Italy
Filippo Furfaro	Università della Calabria, Italy
Massimo Guarascio	CNR-ICAR, Italy
Dragi Kocev	Jozef Stefan Institute, Slovenia
Mirjana Mazuran	Politecnico di Milano, Italy
Ruggero G. Pensa	University of Torino, Italy
Gianvito Pio	University of Bari, Italy
Domenico Potena	Università Politecnica delle Marche, Italy
Jerzy Stefanowski	Poznan University of Technology, Poland
Irina Trubitsyna	University of Calabria, Italy
Herna Viktor	University of Ottawa, Canada
Alicja Wieczorkowska	Polish-Japanese Academy of Information Technology, Poland

Effective Linear Models for Learning with Sequences and Time Series (Abstract of Invited Talk)

Giorgiana Ifrim

University College Dublin, Ireland

Abstract. In this talk I present some of the work done in my research group on developing machine learning algorithms for classification and regression tasks on sequences and time series data. The focus is on algorithms to train linear models. We show that albeit these linear models are considered too simple to achieve high accuracy in many learning tasks, when trained in rich feature spaces they are strong competitors to very complex models such as ensembles and deep learning models. Linear models with rich features are as accurate as complex non-linear models, but are very efficient to train and are interpretable. Interpretability in this context means that the model (a list of weighted features) and the prediction (a sum of feature weights) is transparent to the user. I first provide an overview of important and wide application areas where we encounter sequences and time series, discuss common approaches to learn with sequences, and present algorithms for sequence classification and regression tasks. I also show how ideas from sequence learning can naturally carry over to time series data and show that a linear model with features selected from multiple symbolic representations, achieves state-of-the-art time series classification accuracy. By combining multiple representations of the sequence data to create rich features, we enable linear models to achieve high accuracy, have efficient training and preserve interpretability, the latter being a crucial requirement in many applications.

Contents

Applications

Complex Patterns

A Framework for Pattern Mining and Anomaly Detection in Multi-dimensional Time Series and Event Logs

Len Feremans[1]([✉]), Vincent Vercruyssen[2], Wannes Meert[2], Boris Cule[1], and Bart Goethals[1,3]

[1] University of Antwerp, Antwerp, Belgium
{len.feremans,boris.cule,bart.goethals}@uantwerpen.be
[2] KU Leuven, Leuven, Belgium
{vincent.vercruyssen,wannes.meert}@cs.kuleuven.be
[3] Monash University, Melbourne, Australia

Abstract. In the present-day, sensor data and textual logs are generated by many devices. Analysing these time series data leads to the discovery of interesting patterns and anomalies. In recent years, numerous algorithms have been developed to discover interesting patterns in time series data as well as detect periods of anomalous behaviour. However, these algorithms are challenging to apply in real-world settings. We propose a framework, consisting of generic transformations, that allows to combine state-of-the-art time series representation, pattern mining, and pattern-based anomaly detection algorithms. Using an early- or late integration our framework handles a mix of multi-dimensional continuous series and event logs. In addition, we present an open-source, lightweight, interactive tool that assists both pattern mining and domain experts to select algorithms, specify parameters, and visually inspect the results, while shielding them from the underlying technical complexity of implementing our framework.

1 Introduction

Discovering interesting patterns and anomalous periods in heterogeneous time series data is often the main interest of people generating and analyzing these data. In the past decades, the field of pattern mining has developed a large body of algorithms to automatically discover different types of interesting patterns, such as *frequent itemsets* and *sequential patterns* [21]. However, these algorithms are difficult to use for anyone who is not familiar with their inner workings. Moreover, the algorithms require the data to be preprocessed to the proper format and the type and quality of the patterns being found is largely dependent on the choices made in the preprocessing steps. If a dataset consists of multiple time series or dimensions this becomes even more problematic. Recent algorithms for pattern-based anomaly detection in time series suffer from the same drawbacks [7,11].

© Springer Nature Switzerland AG 2020
M. Ceci et al. (Eds.): NFMCP 2019, LNAI 11948, pp. 3–20, 2020.
https://doi.org/10.1007/978-3-030-48861-1_1

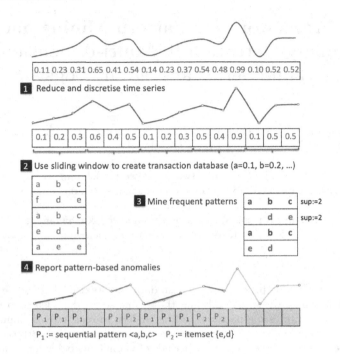

Fig. 1. Example of a pipeline for pattern-based anomaly detection. A window has a high anomaly score if it matches few frequent patterns.

An example pattern-based anomaly detection pipeline is shown in Fig. 1. Here we transform the single continuous signal to a *transaction database* of small discrete sequences using discretisation and a sliding window. Next, we mine frequent patterns in this transaction database and finally give a high anomaly score to windows that match no (or few) frequent patterns. When a new sequential pattern mining or a pattern-based anomaly detection algorithm is presented, important time series representation choices are often only discussed in the experimental design, and a review of alternative representations is often out of scope. Moreover, a wealth of itemset and sequential pattern mining algorithms has been developed in the past decades [8]. Most of these pattern mining algorithms are optimised towards specific, *built-in* constraints, such as mining closed itemsets or mining sequential patterns satisfying temporal constraints [16]. In the literature, little attention is given to generic *external* constraints for reducing the set of discovered patterns independently from any specific algorithm.

In addition, we find that in real-world applications anomalies are often *contextual* [1], that is, an outlier value is only anomalous given that it's out of context. For example, a high temperature during a cold winter night is anomalous, while it's normal during a summer day. Likewise, a shopping bill for more than a thousand dollar is anomalous, except during the Christmas period. Note that we do not require defining contextual attributes, but rather mine patterns of normal behaviour in all input dimensions. In contrast, classic outlier detection algorithms assume a single continuous time series that is stationary, meaning that statistical properties, such as distribution or auto-correlation structure are constant.

We contribute a framework for pattern mining and anomaly detection in time series data. The framework allows its users flexibility regarding the three major steps in the time series analysis workflow: preprocessing, pattern mining, and anomaly detection. First, the framework supports several preprocessing algorithms for representing continuous time series, as well as a generic transformation that creates a transaction or sequence database for both single, multi-dimensional, and mixed continuous and discrete time series data. Second, the framework supports the use of all state-of-the-art pattern mining algorithms for mining itemsets and sequential patterns [8]. In addition, it adds a number of external constraints for reducing the set of discovered patterns independently from any specific algorithm, such as temporal constraints. Third, the framework supports two anomaly detection algorithms [7,11] that are extended to make them compatible with any pattern mining algorithm and multiple dimensions. The framework allows its users to rapidly test various compositions of these three time series analysis building blocks, even new compositions not considered by the original authors of each separate block. For example, instead of frequent sequential patterns, an end-user of our framework can mine a set of sequential patterns using an alternative interestingness measure [6,17], subsequently apply temporal constraints, and then use these patterns as input to an anomaly detection algorithm. Finally we have created an *open-source tool* for Time series Pattern Mining and anomaly detection (TIPM). The tool enables an iterative, exploratory workflow for preprocessing, finding patterns and discovering anomalies, and visualising data and patterns using our framework.

2   Preliminaries

This section clarifies the important time series and pattern mining terminology used throughout the paper. The concepts are largely adapted from [7].

Time Series Data. A *continuous time series* is as a sequence of numerical values $(\langle x_1, t_1 \rangle, \ldots, \langle x_n, t_n \rangle)$, where each real value x_k is associated with a distinct timestamp t_k. A *discrete event log* is a sequence of discrete events $(\langle e_1, t_1 \rangle, \ldots, \langle e_n, t_n \rangle)$ where $e_k \in \Sigma$, with Σ a finite domain of discrete event types. Multiple events can co-occur at the same timestamp. Finally, a *mixed-type time series* $\mathbf{S}$ is a collection of N continuous time series and M event logs and has dimensionality $M + N$. A single time series S^i in $\mathbf{S}$ has only one dimension. It is possible for M or N to be 0. Thus, univariate and multivariate time series are special cases of this definition.

A *time series window* $S^i_{t,l}$ is a contiguous subsequence of a time series S^i containing all measurements for which $\{\langle x_i, t_i \rangle$ or $\langle e_i, t_i \rangle |\ t \leqslant t_i < t + l\}$. A segment of length l can be defined over all dimensions of $\mathbf{S}$ simultaneously.

Pattern Mining. The following definitions are adapted from [21]. An *itemset* X consists of one or more items $x_j \in \Omega$, where Ω is a finite domain of discrete values, that is, $X = \{x_1, \ldots, x_m\} \subseteq 2^{|\Omega|}$. An itemset does not require a temporal order between its items. An itemset X is covered by a window $S_{t,l}^i$ if all items in X occur in that window in any order, denoted as $X \prec S_{t,l}^i$. Given the set of all windows $\mathcal{S}$ of a time series, $cover(X, \mathcal{S})$ is the set of all windows in $\mathcal{S}$ that cover X and $support(X, \mathcal{S})$ is the length of this set.

A *sequential pattern* X_s consists of an ordered list of one or more items, denoted as $X_s = (x_1, \ldots, x_m)$, where $x_j \in \Omega$. A sequential pattern can contain repeating items, and, unlike n-grams, an occurrence of a sequential pattern allows *gaps* between items. A sequential pattern X_s is covered by a window $S_{t,l}^i$ if all items in X occur in that window in the order imposed by the sequential pattern, denoted as $X_s \prec S_{t,l}^i$. The definitions of cover and support are equivalent to those of itemsets. Finally, an itemset or a sequential pattern is *frequent* if its support is higher than a user-defined threshold on *minimal support*.

3 Method

The problem we are trying to solve is defined as follows:

Given: A time series dataset **S** consisting of one or multiple time series.
Do: Find interesting patterns and/or periods of abnormal behaviour in the data.

The general workflow of our framework is shown in Fig. 2. Note that in our framework two strategies are available for finding anomalies. In the first strategy, we create a model of normal behaviour and predict anomalies based on deviations from this model. This is the case for the frequent pattern-based anomaly detection technique, where try to find many patterns that occur frequently and are used for *positive* detection of anomalies. A second strategy is to find anomalous patterns directly or use *negative* detection [4]. Which strategy to use, depends on the use case and can be freely chosen by the user.

3.1   Time Series Representation for Pattern Mining

Dealing with Outliers. If one uses positive detection, it makes sense to remove outlier (extreme) values, that is, cap outlier values that deviate a user-specified number of standard deviations from the mean. If one uses negative detection, it makes sense to keep outlier values and discretise them along with the rest of the data, possibly in a separate bin, as the occurrence of outlier values, is often indicative of contextual anomalies.

Time Series Dimensionality Reduction. A straightforward transformation to reduce time series is piecewise aggregate approximation (PAA) [12]. Given a time series **S**, one sets a window duration l_{PAA} and then replaces each consecutive window in **S** with the mean of the continuous values in the window. This effectively downsamples a time series **S** by a factor $|\mathbf{S}|/l_{PAA}$. In practice, it is often beneficial to downsample each time series as we are more interested in patterns that span a larger period. Note that PAA allows more flexibility than symbolic

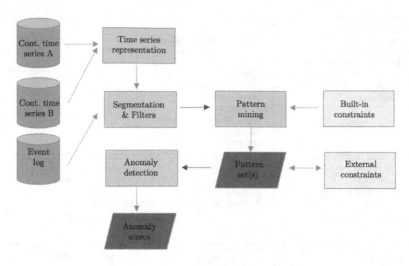

Fig. 2. Workflow of our framework.

aggregate approximation [14]. The latter assumes that the time series values are normally distributed, which is rarely the case in a non-stationary time series.

Discretisation. After reducing dimensionality, the continuous time series are discretised using *equal-width* or *equal-length* bins. As a rule-of-thumb, equal-width discretisation is applied if the observations are normally or uniformly distributed over the bins. If this is not the case, equal-length binning with a slightly larger number of bins can be selected by the end-user. The goal of discretisation is to have good coverage of items that occur in at least 5% of segments.

Segmentation. Before pattern mining, the time series need to be transformed into a *transaction* database. This is done by sliding a fixed-size window over the data and storing each time series window separately as a transaction. The window duration l_{segm} and increment i_{segm} are specified in time units or steps. Setting segmentation parameters is largely domain-specific. For instance, if the length of the datasets is two hours, but measurements (or events) are sampled every second, then finding patterns within 1 min makes sense. The window duration and increment are important parameters towards pattern mining since they directly determine which patterns will found as well as their length. In practice, useful patterns are limited in length so one must ensure that windows are of moderate size by either setting a relatively small value for the duration or by reducing the time series dimensionality.

Filters and Aggregation. Finally, our framework supports basic filtering and aggregation on the time series, as well as generic SQL queries. Filtering is useful if the goal is to model only a part of the dataset. For instance, an end-user can filter the time series on time, on periods where certain warning or error codes occur, or periods where some continuous variable exceeds a certain threshold. This has the advantage that end-users can mine and discover interesting patterns

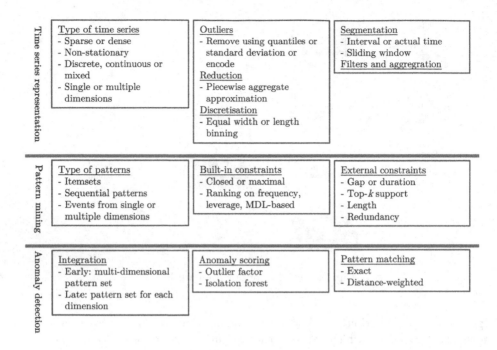

Fig. 3. Detailed overview of our framework.

local to certain events or conditions. Finally, the framework provides options to aggregate values within each window and compute summary statistics such as min, mean, max, count and unique.

Automatically Selecting Parameters. What constitutes a good time series representation depends strongly on the specific application. Good parameters are either selected using domain knowledge or set interactively in a trial-and-error way. However, for the anomaly detection algorithms, it is possible to select parameters using a wrapped approach. Let l_{PAA}, l_{segm}, i_{segm}, and b be the PAA window duration, segmentation length, window increment, and the number of bins respectively. The optimal parameters are selected from the parameter space $\Omega = \{l_{\mathrm{PAA}}, l_{segm}, i_{segm}, b\}$ through optimization of an evaluation metric on the anomaly scores (e.g., AUROC).

3.2   Pattern Mining

After the time series data are discretised and segmented, we can mine patterns. A more detailed overview of our framework is shown in Fig. 3.

Frequent Pattern Mining. An end-user can discover patterns for each dimension of time series **S** that is either discrete or has been transformed into a discrete representation. Our current framework integrates with the SPMF library containing more than 40 algorithms for itemset and sequential pattern mining, covering efficient algorithms for mining frequent, closed, and maximal itemsets and sequential patterns, top-k sequential patterns ranked on leverage and a set of sequential patterns compressed using minimal description length [8,13,17]. For the brevity of this paper, we will not discuss the details of these algorithms and refer to existing work [8,21]. Itemset and sequential pattern mining algorithms require a suitable representation for enumerating patterns and computing support. Itemset mining algorithms require a *transaction database*. This database is created by generating a transaction, or unordered set of *items*, for each window. Likewise, sequential pattern mining algorithms require a *sequence database* where for each window, we create a chronologically ordered list of items (if two events happen at the same time, this is also encoded). Each item is encoded using an integer identifier and either represents an event code or discretised continuous value. We decode item identifiers to report human-readable patterns. An example of maximal itemset mining is shown in Fig. 4.

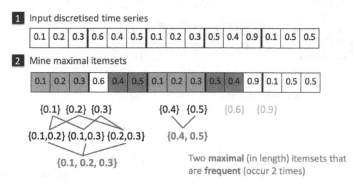

Fig. 4. Example of maximal itemset mining.

External Constraints. A recent benchmark study found that temporal constraints for pattern mining in time series are of high importance [22]. Our framework computes occurrences of itemsets and sequential patterns, reported by any algorithm, and computes the occurrences that have a minimum duration in each window, by looking at the raw dataset. If the minimal occurrence does not satisfy *temporal constraints* on maximal duration and maximal gap (time between two pattern items in one occurrence), we remove the occurrence and re-compute the support for each pattern. In addition, we provide basic external constraints for filtering patterns on the minimum and maximal *length*, filtering the top-k patterns on *support*, and removing *redundant* patterns using a threshold on Jaccard similarity, i.e., if two patterns cover mostly the same transactions, filter out the pattern with the lowest support.

Multi-dimensional Pattern Mining. Thus far, pattern mining algorithms only work on a *single*-dimensional event log or continuous time series, after preprocessing. Our framework makes it possible to uncover patterns with events from multiple dimensions of a time series **S**. Under this *early integration* strategy, a transaction (or sequence) database is created from the time series by adding events from multiple dimensions of the time series to a transaction. Similarly, sequence transactions (necessary for sequential pattern mining) are created by adding events from multiple dimensions in a chronological fashion. We differentiate between events from different dimensions by encoding the item identifier to reflect the source dimension. An example pipeline is shown in Fig. 5.

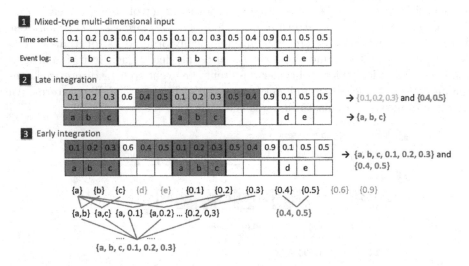

Fig. 5. Example of mining pattern in multi-dimensional time series. Under the early integration strategy, events from both dimensions are considered simultaneously. Under the late integration strategy, patterns are mined in each dimension separately.

Pattern Explosion in Time Series. While the pattern mining community has gone through great lengths in creating efficient algorithms for different tasks, time series remain a difficult data source for efficient pattern mining. For example, imagine a time series that contains a sequence of 20 values and occurs frequently. Because this series is frequent, any subsequence will also be frequent, thereby generating an exponential number of patterns. In general, time series generate a lot of patterns due to naturally occurring autocorrelation. The problem becomes even worse when two or more dimensions are added, especially if different time series dimensions are highly correlated. In practice, we prefer mining maximal patterns with relatively high support *in each dimension separately*. This strategy, dubbed the *late integration* strategy, is illustrated in Fig. 5. Alternatively, we can change the representation of the time series. In our experience, we find that using pattern sets with more than a few thousand of patterns rarely results in higher accuracy.

3.3   Pattern-Based Anomaly Detection

Our framework supports two algorithms for anomaly detection: a generalised version of frequent pattern-based outlier factor (FPOF) and a generalised version of pattern-based anomaly detection (PBAD) [7,11]. Both methods take a set (or sets) of patterns as input and compute an anomaly score between 0.0 (normal) and 1.0 (abnormal) for each time series segment. By setting the window increment i_{segm} equal to a single time step, it is possible to compute the anomaly score at each timestamp. Figure 6 shows an example of both anomaly detection approaches.

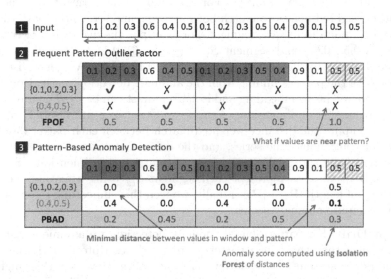

Fig. 6. Example of FPOF and PBAD for computing anomalies based on a previously discovered pattern set. In FPOF the anomaly score is based on the number of exactly matching patterns. In PBAD we compute the distance between each window and pattern and compute scores using an isolation forest.

Generic Outlier Factor. FPOF [11] computes an anomaly score a for each segment $S_{t,l}^i$ in time series $\mathbf{S}$, given a pattern set $\mathbf{P}$, based on the total number of patterns matching each segment, denoted by $P_k < S_i$:

$$a(S_{t,l}^i, \mathbf{P}) = 1.0 - \frac{|\{P_k | P_k \in \mathbf{P} \text{ and } P_k < S_{t,l}^i\}|}{|\mathbf{P}|}.$$

The authors only consider closed itemsets over a single dimension, but we can extend FPOF to compute this score for any pattern set, such as sequential patterns, and for multiple pattern sets mined over multiple dimensions of $\mathbf{S}$. Given two patterns sets, $\mathbf{P}_1$ and $\mathbf{P}_2$, the anomaly score is computed as:

$$a(S_{t,l}^i, \mathbf{P}_1 \cup \mathbf{P}_2) = 1.0 - \frac{|\{P_k | P_k \in \mathbf{P}_1 \cup \mathbf{P}_2 \text{ and } P_k < S_{t,l}^i\}|}{|\mathbf{P}_1 \cup \mathbf{P}_2|}.$$

It is trivial to extend this formula to d dimensions. The only requirement is that for computing a match from dimension d, we need to check if the pattern mined from dimension d matches the segment of the corresponding dimension. Multiple pattern sets can also be mined over the same dimension using a different algorithm or settings. For example, we can mine both itemsets and sequential patterns in a single dimension S^i.

Generic Isolation Forest of Distance-Weighted Occurrences. PBAD [7] computes anomaly scores with the help of the isolation forest algorithm applied to an embedding of both maximal itemsets and sequential patterns for each continuous and discrete dimension [7]. For continuous time series, the authors use a distance-weighted match to match both itemsets and sequential patterns with each original, non-discretised, segment. For example, the distance between itemset $P_k = \{'0.5', '0.6'\}$ and segment $S_1 = (\underline{0.50}, \underline{0.61}, 0.11, 0.10)$ will be smaller than the distance to segment $S_2 = (\underline{0.31}, \underline{0.42}, 0.12, 0.04)$. We generalise PBAD by decoupling the pattern mining from the anomaly detection phase. Concretely, the distance-weighted embedding and isolation forest can be used on any pattern set and any number of dimensions. Assume we have two pattern sets $\mathbf{P}_1$ and $\mathbf{P}_2$. First, we compute the distance-weighted match between each pattern and each window for continuous time series, and the exact match for discrete (or multi-dimensional) time series. We now have two matrices of dimensions $|\mathbf{S}| \times |\mathbf{P}_1|$ and $|\mathbf{S}| \times |\mathbf{P}_2|$, and can represent each segment $S_{t,l}^i$ using a feature vector (or embedding) of length $|\mathbf{P}_1| + |\mathbf{P}_2|$. Finally, we feed this embedding to an isolation forest to compute anomaly scores.

Concept Drift. For pattern-based anomaly detection, we assume a stable distribution such that the mined patterns are good descriptors of the new data that enters the system and deviations are anomalies. However, this might not be true in practice, especially over a long period of time where the observed system might change. In such a setting we can use the pattern-based anomaly detection as part of an online adaptive learning procedure [9] and extend our framework to detect *concept drift*. The anomaly score is, in this case, the target variable that is being predicted from the new instances, and the loss function is the deviation from an average anomaly score closer to 0, representing normal behaviour. When the aggregated loss grows too large or some other change point detection algorithms crosses a threshold, the framework signals concept drift. Depending on the application, various strategies can be used to relearn. From maintaining a database of previous data to gradual forgetting old patterns and introducing new mined patterns to the pattern set(s). We refer to Gama et al. for an extensive overview.

3.4   Implementation of the Framework

We implemented our framework in Java as an open-source web-based application called TIPM[1].

---

[1] Source and datasets available at https://bitbucket.org/len_feremans/tipm_pub.

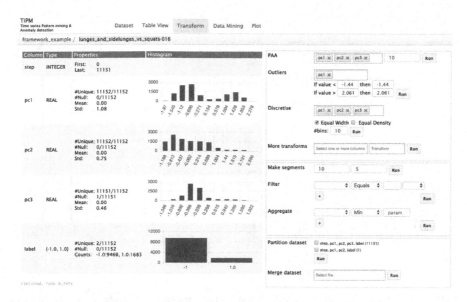

Fig. 7. TIPM: Time series representation options. In the first use case, we apply PAA with $l_{PAA} = 10$, cap outlier values, discretise all time series using 16 equal width bins and create overlapping segments with a $l_{segm} = 10$ and $i_{segm} = 5$.

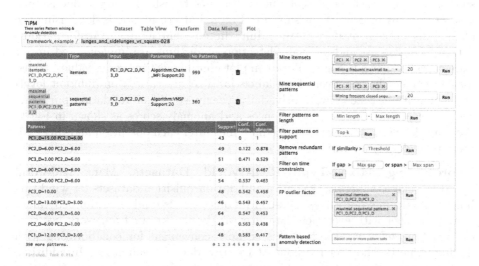

Fig. 8. TIPM: Using an early integration strategy, we mined maximal itemsets and maximal sequential patterns with minimal $support = 20$ and compute the generalised FPOF anomaly score. An example of an anomalous pattern is highlighted.

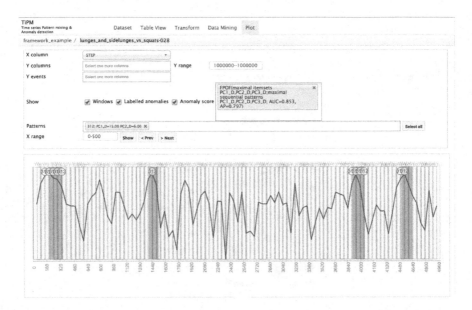

Fig. 9. TIPM: Visualisation of anomaly scores and patterns. We show occurrences of the interpretable anomalous pattern $pc1 = 15$, $pc2 = 6$.

Interactive Workflow. TIPM takes in any dataset that contains at least a timestamp and one or more value columns. TIPM visualises the histogram and summary statistics for each column and allows transforming continuous time series using our framework, as shown in Fig. 7. Pattern mining is done using the algorithms implemented in SPMF. Multi-dimensional mining transformations, external constraints and anomaly detection algorithms are implemented in our framework as shown in Fig. 8. TIPM can plot continuous time series values, transformed values, discrete event logs, labels, and segmentation, on different levels of granularity in time (raw, hourly, daily, yearly, etc.). For validation purposes, the tool can render pattern occurrences and anomaly scores as shown in Fig. 9. We remark that TIPM saves intermediate files after each operation allowing end-users to undo any operation.

Representing Mixed-Type Real-World Datasets. Many real-world datasets, such as supervisory control and data acquisition datasets for wind turbines, contain missing values, non-continuous periods, and timestamped values stored together with event log data in a single file. In our framework, we stay close to this *tabular format* as this is most convenient for collaborating with domain experts who prefer to look at the raw data for validation. In addition, we provide two explicit temporal join operations: *partition* and *merge*. Partition takes a subgroup of columns having non-zero values and saves them in a separate table. This is useful for extracting event log data from continuous time series data. Merge is the opposite operation and takes the union of two tables and sorts them on time. If two column names match in both tables, merge takes the column value of the first table. For example, merge can be used to join time series datasets from multiple devices.

Scaling to Large Datasets. In our implementation, we use *streaming* operations as much as possible. Each procedure implemented in this way, processes one row at a time, instead of loading all data into main memory. Alternative, we load data in a *paginated* way, that is, we only load data required in the user interface, i.e. only the current period of the time series. By processing data using streaming operations and loading data paginated, the interface and pre- and postprocessing transformations can handle large time series with millions of samples instantaneously. For pattern mining, we can manage resources by setting support to a relatively high value, and reducing time series as discussed before. A possible extension would be to include streaming pattern mining algorithms.

4   Use Cases

In this section, we will illustrate the usefulness of our framework, implemented in TIPM, using two use cases.

Anomaly Detection on Multivariate Times Series. For the first use case, we detect anomalies in a multivariate time series dataset that was obtained by using a Kinect sensor to track the body movements during indoor physical exercises [2]. The goal is to assist people in performing the exercises correctly. We focus on detecting incorrectly executed exercises during a continuous workout session consisting of 60 lunges and 10 squats. The ground-truth values are known. The original dataset consists of 75 time series and we reduced this to 3 time series using principal component analysis [7].

First, we upload the time series in tabular file format. TIPM shows statistics and histograms for each time series (*pc1*, *pc2* and *pc3*) as well as the label as shown in Fig. 7. We can now select options to preprocess each time series. First, we cap outlier values based on the 1% and 99% quantiles. Next, we compute and store the average value every 10 time steps, that is $l_{PAA} = 10$, to reduce the 3 continuous dimensions using PAA. We then apply equal-width discretisation with $b = 16$ bins. For multi-dimensional mining, our input dataset thus consists of 16×3 discrete items. We create sliding windows with a duration of $l_{segm} = 10$ (3 seconds in absolute time) and $i_{segm} = 5$, resulting in 223 windows of length 10 that overlap for 50%. With all continuous data time series represented as discretised segments, we start mining patterns. We opt for early integration and select all three dimensions as input. We select an algorithm for mining maximal itemsets (CHARM_MFI) with a minimum *support* of 20%. For reducing patterns, we remove itemsets that co-occur in at least 90% of windows, resulting in 999 itemsets. We compare this set of patterns by mining maximal sequential patterns (VMSP) with the same settings, resulting in 360 patterns. Finally, we run the generic outlier factor and compute an anomaly score for both types of pattern sets individually. The screens in TIPM are shown in Fig. 8 for mining and Fig. 9 for visualisation. We show occurrences of the (anomalous) pattern

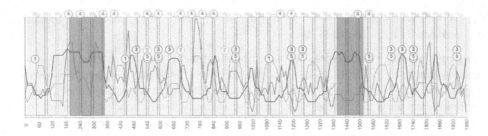

Fig. 10. Visualisation in TiPM of the first minute of data. The blue line is *pc1*, the light blue line *pc2*, and the orange line *pc3*. We show the top-5 most frequent maximal itemsets, mined over all three dimensions. (Color figure online)

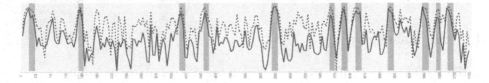

Fig. 11. Visualisation in TiPM of anomaly detection results. Segments with a red background are labelled anomalies, and the black line is the anomaly score predicted (unsupervised) using generic outlier factor using maximal sequential patterns. The dotted line is the results using maximal itemset patterns. (Color figure online)

$pc1 = 15, pc2 = 6$, found by sorting all maximal sequential patterns with minimal support of 20 on decreasing confidence towards labelled anomalies. Figure 10 shows the first minute of the Kinect dataset. TiPM shows the transformed time series and overlapping segments. We selected the top-5 most frequent maximal itemsets for visualisation. The first itemset is $\{pc1 = 1, 2, 4 \wedge pc2 = 1 \wedge pc3 = 4, 5\}$ which has a support of 56 (or relative support of 0.25). This means that 25% of segments contain both (discretised) values of 1, 2 and 4 in time series *pc1*, 1 in *pc2*, and 4 and 5 in *pc3*. Notice that the first frequent pattern, as well as the 2nd, 3th and 5th, but not 4th, almost never occur in any anomalous segment highlighted in red. Consequently, the patterns are examples of frequent interpretable patterns that occur during normal behaviour. Deviations from these patterns are marked as anomalies. As mentioned before, TiPM allows sorting patterns on confidence towards normal or abnormal segments, thereby assuming labels. We find that sequential patterns containing high values of *pc1* are the most predictive towards abnormal behaviour. Figure 11 shows the anomaly scores over the entire 6 min time series. Using the generic outlier factor anomaly detection method we can report an AUROC of 0.839 and average precision of 0.767 for maximal itemsets, and an AUROC of 0.884 and average precision of 0.833 for maximal sequential patterns.

Exploratory Analysis of Real-World Heterogeneous Time Series. In the second use case, we perform an exploratory analysis of a supervisory control and

Fig. 12. TIPM: Use case for exploratory analysis of heterogeneous wind turbine data. We show two interesting patterns that are characteristic of operational behaviour, mined from the event log.

data acquisition dataset collected from a wind turbine farm [5]. This dataset is challenging because: (i) the data was collected over different years, (ii) there are multiple continuous time series, (iii) there is an event log containing more than hundreds of different types of events, (iv) behaviour of a wind turbine is strongly dependent on current weather conditions. Figure 12 shows the preprocessed data and two interesting patterns for a selected period of 1 month in TIPM. For the wind turbine, we have selected two continuous variables representing the wind speed and power output as well as an event log containing warning, error, and operational codes. First, we normalised both continuous time series. We verify that wind speed and power output are highly correlated. The main difference is that power output is capped to a maximal value. The different occurring events are shown as colour-coded dots. We mine maximal itemsets with a minimal support of 1 and found about 140 patterns. Next, we filter patterns with a minimum size of 2. Next, we remove redundant patterns by setting a Jaccard similarity threshold of 0.9, thereby removing patterns that co-occur in 90% of windows. From the remaining patterns, we show two maximal itemsets of size 5 and 6 that occur in the selected period. Both patterns correspond to a specific series of operator actions for remotely stopping and restarting the turbine. If these patterns occur, the power output drops to 0, regardless of the current wind speed. From this second use case, we conclude that our framework can be used to explore complex multi-dimensional datasets, using patterns extracted from the event log, to capture meaningful operational behaviour.

5 Related Work

Most general data mining libraries, such as WEKA or KNIME, are incomplete concerning pattern mining. TIPM is complementary to SPMF [8] by implementing temporal constraints, multi-dimensional pattern mining, and pattern-based anomaly detection algorithms. In contrast to SPMF, and other libraries that implement time series transformations on consecutive numeric vectors, we support timestamped tabular data with multiple dimensions, and mixed-type attributes. Other tools for anomaly detection in time series uses either shapelets or motifs (or discords) in single-dimensional continuous time series [18]. Interactive pattern mining tools, such as MIME [10] or SNIPER [15], do not support continuous time series.

There exist algorithms for directly mining patterns with temporal constraints [16]. However, by providing temporal constraints as an external post-processing filter, we can apply them to any pattern mining algorithm. This is of interest for many efficient algorithms for mining closed, maximal or interesting patterns that do not support temporal constraints. Many more transformations for reducing the length of the time series exist [3]. We prefer PAA for two reasons. First, different authors have confirmed that more advanced techniques are not necessarily more effective [3,14]. Second, many other representation techniques, i.e., transformation to spectral space, single value decomposition, or clustering, make interpretation much harder while patterns of binned values are easy to interpret. Remark that other transformations, such as differencing or smoothing the raw time series are not problematic regarding interpretation.

Two popular techniques for classification and anomaly detection in time series are the *matrix profile* [20], that computes an outlier score relative to the euclidean or dynamic time warping (DTW) distance to its nearest neighbour, and time series *shapelets*, which are subsequences from a continuous time series and are used in combination with the DTW distance to classify time series segments [19]. A key difference is that frequent patterns naturally handle both continuous time series and event logs. If we compare sequential patterns to shapelets, we argue that on the one hand, sequential patterns generalise shapelets, because we use non-continuous subsequences with gaps. On the other hand, sequential patterns are more specific, because they consist of discretised values instead of continuous values. The latter argument against sequential patterns, however, can be relaxed by using a weighted distance. Itemsets, however, are radically different from shapelets and of value for predicting anomalies. In future work, an *ensemble* of representations could have value. That is, we can compute itemset and sequential pattern distances, exact pattern matches, shapelet distances, motif distances, and combine those in one feature vector, as input for existing classification or anomaly detection algorithms.

6 Conclusion

Existing pattern-based anomaly detection algorithms focus on a particular combination of time series representation, pattern mining, and computation of the anomaly score. In PBAD, the authors remarked that this method is a promising general framework for time series anomaly detection, where certain variations might be more effective in different applications [7]. In this paper, we implement such a framework and discuss a wealth of general building blocks, that can be composed to create new variations. This allows data scientists to create novel unsupervised anomaly detection models. We also present TIPM, an interactive, easy-to-use, and open-source tool that implements our framework. TIPM is unique since we have a rich set of options for interactively preprocessing and mining patterns from mixed-type time series, supported by visualisation of (raw and transformed) time series, event logs, segments, patterns and anomaly scores. With our framework, we show how to discover interesting interpretable patterns and detect anomalies in multi-dimensional time series in two different use cases.

Our framework and corresponding tool are designed to support real-world applications. For applications such as condition monitoring of devices, it is important to support devices that log both sensor values and events. We focus on contextual anomalies, i.e. we only consider outlier values as anomalous if they are abnormal given the current operational conditions, by capturing normal behaviour using patterns and predicting anomalies as deviations from normal behaviour. We also discussed the integration of concept drift within our framework as an important next step.

Acknowledgements. The authors would like to thank the VLAIO SBO HYMOP project for funding this research.

References

1. Chandola, V., Banerjee, A., Kumar, V.: Anomaly detection: a survey. ACM Comput. Surv. (CSUR) **41**(3), 15 (2009)
2. Decroos, T., Schütte, K., De Beéck, T.O., Vanwanseele, B., Davis, J.: AMIE: automatic monitoring of indoor exercises. In: Brefeld, U., et al. (eds.) ECML PKDD 2018. LNCS (LNAI), vol. 11053, pp. 424–439. Springer, Cham (2019). https://doi.org/10.1007/978-3-030-10997-4_26
3. Ding, H., Trajcevski, G., Scheuermann, P., Wang, X., Keogh, E.: Querying and mining of time series data: experimental comparison of representations and distance measures. Proc. VLDB Endow. **1**(2), 1542–1552 (2008)
4. Esponda, F., Forrest, S., Helman, P.: A formal framework for positive and negative detection schemes. IEEE Trans. Syst. Man Cybern. Part B (Cybern.) **34**(1), 357–373 (2004)
5. Feremans, L., Cule, B., Devriendt, C., Goethals, B., Helsen, J.: Pattern mining for learning typical turbine response during dynamic wind turbine events. In: ASME 2017 International Design Engineering Technical Conferences and Computers and Information in Engineering Conference, p. V001T02A018. American Society of Mechanical Engineers (2017)
6. Feremans, L., Cule, B., Goethals, B.: Mining top-k quantile-based cohesive sequential patterns. In: Proceedings of the 2018 SIAM International Conference on Data Mining, pp. 90–98. SIAM (2018)

7. Feremans, L., Vercruyssen, V., Cule, B., Meert, W., Goethals, B.: Pattern-based anomaly detection in mixed-type time series. In: Joint European Conference on Machine Learning and Knowledge Discovery in Databases (2019)
8. Fournier-Viger, P., et al.: The SPMF open-source data mining library version 2. In: Berendt, B., et al. (eds.) ECML PKDD 2016. LNCS (LNAI), vol. 9853, pp. 36–40. Springer, Cham (2016). https://doi.org/10.1007/978-3-319-46131-1_8
9. Gama, J., Žliobaitė, I., Bifet, A., Pechenizkiy, M., Bouchachia, A.: A survey on concept drift adaptation. ACM Comput. Surv. (CSUR) **46**(4), 44 (2014)
10. Goethals, B., Moens, S., Vreeken, J.: Mime: a framework for interactive visual pattern mining. In: Proceedings of the 17th ACM SIGKDD International Conference on Knowledge Discovery and Data Mining, pp. 757–760. ACM (2011)
11. He, Z., Xu, X., Huang, Z.J., Deng, S.: FP-outlier: frequent pattern based outlier detection. Comput. Sci. Inf. Syst. **2**(1), 103–118 (2005)
12. Keogh, E., Chakrabarti, K., Pazzani, M., Mehrotra, S.: Dimensionality reduction for fast similarity search in large time series databases. Knowl. Inf. Syst. **3**(3), 263–286 (2001)
13. Lam, H.T., Mörchen, F., Fradkin, D., Calders, T.: Mining compressing sequential patterns. Stat. Anal. Data Mining: ASA Data Sci. J. **7**(1), 34–52 (2014)
14. Lin, J., Keogh, E., Lonardi, S., Chiu, B.: A symbolic representation of time series, with implications for streaming algorithms. In: Proceedings of the 8th ACM SIGMOD Workshop on Research Issues in Data Mining and Knowledge Discovery, pp. 2–11. ACM (2003)
15. Moens, S., Jeunen, O., Goethals, B.: Interactive evaluation of recommender systems with sniper - an episode mining approach. In: Proceedings of Thirteenth ACM Conference on Recommender Systems. RecSys 2019, September 2019
16. Pei, J., Han, J., Wang, W.: Constraint-based sequential pattern mining: the pattern-growth methods. J. Intell. Inf. Syst. **28**(2), 133–160 (2007)
17. Petitjean, F., Li, T., Tatti, N., Webb, G.I.: Skopus: mining top-k sequential patterns under leverage. Data Mining Knowl. Discov. **30**(5), 1086–1111 (2016)
18. Senin, P., et al.: GrammarViz 2.0: a tool for grammar-based pattern discovery in time series. In: Calders, T., Esposito, F., Hüllermeier, E., Meo, R. (eds.) ECML PKDD 2014. LNCS (LNAI), vol. 8726, pp. 468–472. Springer, Heidelberg (2014). https://doi.org/10.1007/978-3-662-44845-8_37
19. Ye, L., Keogh, E.: Time series shapelets: a new primitive for data mining. In: Proceedings of the 15th ACM SIGKDD International Conference on Knowledge Discovery and Data Mining, pp. 947–956. ACM (2009)
20. Yeh, C.C.M., et al.: Matrix profile i: all pairs similarity joins for time series: a unifying view that includes motifs, discords and shapelets. In: 2016 IEEE 16th International Conference on Data Mining (ICDM), pp. 1317–1322. IEEE (2016)
21. Zaki, M.J., Meira, W.: Data Mining and Analysis: Fundamental Concepts and Algorithms. Cambridge University Press, Cambridge (2014)
22. Zimmermann, A.: Understanding episode mining techniques: benchmarking on diverse, realistic, artificial data. Intell. Data Anal. **18**(5), 761–791 (2014)

A Heuristic Approach for Sensitive Pattern Hiding with Improved Data Quality

Shalini Jangra$^{(\boxtimes)}$ and Durga Toshniowal

Indian Institute of Technology Roorkee, Roorkee 247667, India
shalinijangra312@gmail.com, durgatoshniwal@gmail.com

Abstract. Frequent itemset mining can be used to discover various interesting patterns present in dataset. However, this imposes a great privacy threat when data is shared with other organisations. There are some business critical frequent patterns that are considered as sensitive from organization's or individual's perspective because revealing such patterns can disclose confidential information. Privacy preserving data mining (PPDM) provides various techniques to hide sensitive patterns to make sure that they cannot be revealed by applying data mining models on shared datasets. Heuristic based sensitive pattern hiding techniques are widely adopted PPDM techniques due to their fast execution time but causes high side effects. In this paper, we propose a heuristic approach for sensitive pattern hiding based on deletion of Victim items which is named MinMax. In the proposed algorithm, Misses Cost Impact (*MCI*) value of each tentative Victim item is calculated and item with minimum *MCI* is selected as Victim item resulting in low Misses Cost. Experimental results on benchmark datasets show that proposed algorithm achieves better data quality with less execution time as compared to existing heuristic based techniques.

Keywords: Privacy preserving data mining · Data privacy · Sensitive patterns · Hiding Failure · Misses Cost

1 Introduction

Frequent itemset mining to discover unrevealed patterns present in data benefits the businesses in their various decision making policies. Unveiling of these hidden patterns brings a threat to the privacy of sensitive and confidential information present in data [1]. For example, analysis of financial and medical records can give remarkable business and research benefits but privacy breach might allow business competitors and malicious users to misapply the information that can incur great remunerative and social loss. The confidential knowledge present in the data can be inferred by the frequent patterns called sensitive patterns which can be generated by applying data mining models. These patterns are generally

M. Ceci et al. (Eds.): NFMCP 2019, LNAI 11948, pp. 21–35, 2020.
https://doi.org/10.1007/978-3-030-48861-1_2

given by the user or organisation. Privacy preserving data mining was introduced to diminish privacy issues by concealing the sensitive information while enabling data mining models to extract required information. A large number of sensitive pattern hiding techniques are proposed by various researchers, which are majorly divided into three categories i.e. heuristic based, border based and exact techniques.

Heuristic based techniques have drawn more attention of researchers due to their simplicity and fast execution time [16,17]. However, these techniques experience high side effects and provide suboptimal solution [22]. Maintaining the adequate balance between data quality and data privacy is the prominent issue for sensitive pattern hiding algorithms because if no required information can be mined from the data, there is no use of hiding all the sensitive information. The quality of any sensitive pattern hiding technique predominantly depends on two performance metrics: Misses Cost (MC) and Hiding Failure (HF). Number of non-sensitive frequent patterns accidently concealed in order to conceal sensitive patterns accounts for Misses Cost. Number of sensitive frequent patterns that are not concealed by pattern hiding technique accounts for Hiding Failure. These two factors clearly depend on two things: *Victim Item Selection* and *Transaction Selection*. Many heuristic techniques [3,11] removes some of sensitive transactions from dataset to decrease the support count of sensitive itemsets below minimum support threshold that can result in great reduction of dataset size. Support of an itemset is equal to the fraction of transactions having that itemset with respect to total dataset size. Some of the techniques are based on the deletion of the Victim items from sensitive transactions. Most of Victim item deletion techniques select the item on the basis of support count. For instance, MaxFIA [16] selects the item having highest support count as Victim item since it results in less probability of non-sensitive itemsets to be infrequent. Also selection of optimal transaction to delete Victim item plays a crucial role which can on the basis of transaction size [21], degree of conflict (DoC) [16], relevance of transaction with non-sensitive information (RoT) [5], etc. Now-a-days, metaheuristic based algorithms that used evolutionary approaches like genetic algorithm [12,13], particle swarm optimization [14], ant colony system [25], etc. are also gaining interest of researchers. However, these approaches take high execution time due to hundreds of iterations performed in search of optimal solution. Further, the results of these approaches are highly dependent on the parameters specified by the user such as mutation rate, chromosome size, population size, etc.

Border based techniques [15,23] are based on the border theory in which search space for calculating the side effects on the non-sensitive patterns is reduced to the elements present at the border of the lattice of frequent and infrequent itemsets. These techniques have less side effects as compared to heuristic based approaches but have high computational complexity. Shivani et al. [22] proposed a border based algorithm which uses the MapReduce framework for parallel processing of transactions to sanitize the data. Exact techniques [7,8] consider the problem of sensitive pattern hiding as constraint satisfaction problem and find the optimal solution using linear programming. These techniques are very slow compared to heuristic based techniques.

This paper proposes a new heuristic based algorithm, **MinMax** that differs in the method of *Victim Item Selection*. The principle behind MinMax algorithm is to select the Victim item which appears in less number of non-sensitive patterns and first mask the sensitive itemset whose Victim item appears in more number of non-sensitive itemsets as compared to Victim items of other sensitive itemsets. The concept of *RoT* is chosen for selecting the transaction. Many experiments are conducted on some benchmark datasets to compare the performance of the proposed algorithm with some traditional heuristic techniques which demonstrate that MaxMin preserves a better data quality with commensurate execution time.

The remainder of this paper is organised as follows. Section 2 presents a briefing on some existing sensitive patterns hiding heuristic techniques. Section 3 provides a brief introduction of basic terminologies and the problem statement. Section 4 describes proposed algorithm with demonstrating example. In Sect. 5, the performance of the proposed algorithm is analyzed with experiments. Final conclusion is given in Sect. 6.

2 Related Work: Heuristic Based Algorithms

Numerous heuristic based sensitive pattern hiding and association rule hiding techniques exist in the literature that sanitize the data before applying data mining models. The authors in paper [4] proposed a sensitive rule hiding approach in which sensitive itemset with highest support is preferred to hide first. In this paper, a graph of large itemsets is formed. This itemset graph is traversed bottom up followed by top down to mask sensitive patterns. This approach works fine on small datasets but not feasible on large datasets. Apart from hiding all sensitive patterns, other performance factors of the algorithm are not evaluated. In paper [16], three itemset hiding heuristics named MaxFIA, MinFIA and IGA were proposed to maintain better data quality with data privacy. Transactions are selected on the basis of their degree of conflict. An efficient, scalable and one-scan heuristic named Sliding Window Algorithm (SWA) is proposed in the paper [17]. Transactions are selected in increasing order of their length since shorter transactions have less combinations of sensitive rules. Highest frequency item of sensitive itemsets present in the selected transaction is chosen as victim item. Transactions coming under a k-sized window are sanitized once in sequential manner that imposes a scalability issue for large datasets. The paper [3] presents three heuristic approaches (*Aggregate*, *Disaggregate* and *Hybrid*) that promise better data quality at the cost of computational speed. *Aggregate* approach is based on transaction deletion while *Disaggregate* approach alter the transaction by deleting the items to reduce support of sensitive itemsets. *Hybrid* approach is a combination of the above two approaches that first identifies the transaction using *Aggregate* approach and then deletes item from the selected transaction using *Disaggregate* approach.

Some researchers proposed blocking based rule hiding approaches that replace sensitive information by some unknown items [20, 21, 24]. In the blocking approach, apart from the addition of unknown items, the rest of the dataset remains the same. Therefore it becomes easy to restore the original dataset [19]. In paper [5], the item with maximum support count is selected as Victim Item like MaxFIA but transactions are selected in descending order of their relevance with non-sensitive itemsets. The paper [26] proposed an efficient distortion based rule hiding method through deletion and reinsertion of items. To reduce the Misses Cost, correlation between sensitive and non-sensitive rule is calculated and item with minimum influence on non-sensitive itemsets is selected as Victim item. The papers [6] and [9] proposes sanitization methods on incremental datasets. Paper [6] maintains a tree like data structure to enhance the execution speed and have less side effects. Although it is not efficient for dense datasets. In the approach, proposed in [9], the sanitization process is applied only on the incremented part of the dataset. A dynamic itemset hiding algorithm that considers the multiple support threshold values ie proposed in the paper [18]. It uses item-deletion based sanitization approach on the whole dataset hence reduces Misses Cost.

Above discussion ensures that there is great scope for researchers to explore different aspects of PPDM techniques like quality and scalability on different types of datasets i.e. static datasets, incremental datasets, dense datasets and sparse datasets etc.

3 Background

This section provides a brief introduction of basic terminologies and the problem statement.

3.1 Basic Terminologies

1. **Frequent Itemsets:** Any itemset f_i having support greater than minimum support threshold value is frequent itemset i.e. $\sup(f_i) \geq |D| \times \delta$, where $\sup(f_i)$ is equal to the total number of transactions having itemset f_i, $|D|$ is the size of dataset D and δ is minimum support threshold value. For example, Table 3 shows the discovered frequent itemsets of an example database shown in Table 1 under $\delta = 0.5$.
2. **Sensitive Itemsets:** If the presence of any frequent itemset is able to discover any sensitive pattern, sequence etc. that can reveal some personal and confidential information regarding a company or an individual which they don't want to share, then it will be considered as a sensitive itemset. For example, any attribute or combination of attributes that can reveal the identity of a patient in medical records is considered as sensitive.
3. **Degree of Conflict (DoF):** It is defined as the number of sensitive itemsets a transaction T contains. If $S = \{s_1, s_2, ..., s_n\}$ is set of sensitive itemsets, then

$$DoF(T) = \sum_{i=1}^{n} T(s_i) \tag{1}$$

Table 1. An example database   **Table 2.** Projected database

TID	Items
T1	B C D E F
T2	A C E F G
T3	C D F
T4	A C F G
T5	B C D F G
T6	C E G

TID	Items
T1	B C D E F
T2	A C E F G
T3	C D F
T4	A C F G
T5	B C D F G
T6	C E G

where $T(S_i) = 1$, when $S_i \subseteq T$, otherwise $T(S_i) = 0$.

4. **Relevance of Transaction (RoT):** The relevance of a transaction is calculated as:

$$RoT(T) = \frac{1}{1 + NUM_{non-sens}(T)} \tag{2}$$

where $NUM_{non-sens}(T)$ is equal to the number of non-sensitive itemsets transaction T supports.

5. **Misses Cost Impact (MCI):** Misses Cost Impact of an item 'i' equal to the total number of non-sensitive itemsets in which item 'i' appears.

6. **Victim Item:** Victim item x of a sensitive itemset s_i is an 1-itemset, such that $x \subseteq s_i$ and x is chosen for deletion in order to mask s_i.

3.2 Problem Statement

The problem of sensitive pattern hiding is described as follows. Let D be the original source dataset and F is set of associated frequent itemsets under some minimum support threshold say δ. Let S is the subset of a set of frequent itemsets having itemsets that can be helpful to derive confidential patterns hence considered as set of sensitive itemsets. The problem of sensitive frequent itemset hiding is to sanitize the data by decreasing the support count of sensitive frequent itemsets less than minimum support count value i.e $\sup(s_i) < |\ D\ | \times \delta$, so

Table 3. Discovered frequent itemsets

1-itemset	Ccount	2-itemset	Count	3-itemset	Count
C	6	FG	3	CFG	3
D	3	CE	3	CDF	3
E	3	CD	3		
F	5	CG	4		
G	4	DF	3		
		CF	5		

that sensitive itemsets do not appear as frequent itemsets in sanitized dataset. The problem of sensitive patten hiding mainly revolves around two things: 1) which item should be selected as Victim item for deletion to suppress a particular pattern, 2) From which transaction that selected Victim item should be deleted. Removal of Victim items results into hiding of non-sensitive frequent patterns that accounts for increasing Misses Cost. Therefore, items having less impact on non sensitive itemsets should be selected as Victim items. Transactions supporting at least one of sensitive itemsets should be considered for modification, since alteration of other transactions does not exert any impact on support of sensitive patterns. Therefore, the process of sensitive pattern hiding is transforming of original dataset D into released dataset D' such that most of non-sensitive information and none of sensitive information can be derived from D'.

4   Proposed Solution: MinMax Algorithm

The rationale behind MinMax algorithm is to select the item which appears in less number of non-sensitive patterns as Victim item. Misses Cost Impact (MCI) of an item gives the count of the number of non-sensitive itemsets in which that item appears. MCI of each 1-frequent itemsets is calculated using Algorithm 1. A list called Affinity List (AL) is maintained to have tentative Victim items and corresponding MCI values. After calculating MCI values, dataset is sanitized using Algorithm 2. For each sensitive itemset, the item having the lowest MCI value as Victim item (step 1–3) is selected, that contributes to Min part of algorithm MinMax. The item having the lowest MCI value is selected because it will reduce Misses Cost. If there is more than one item having the lowest MCI value, go for the item which appears in more number of sensitive patterns. Then, sensitive itemsets are sorted in decreasing order of their Victim item's MCI value (step 4) such that sensitive itemset whose Victim item's MCI value is largest as compared to MCI values of other sensitive itemset's Victim items

---

Algorithm 1. MCI Calculation

Input: Set of 1-frequent itemsets, set of non-sensitive itemsets i.e NS
Output: Affinity list with items and corresponding MCI

1: Create an Affinity list AL having tentative victim item and corresponding MCI value.
2: **for** each 1-frequent item x **do**
3:     $MCI(\text{x})=0$
4:     **for** each $ns_i \in$ NS **do**
5:         **if** x $\subseteq NS_i$ **then**
6:             Increment $MCI(\text{x})$ by 1.
7:         **end if**
8:     **end for**
9:     AL.append(x, $MCI(\text{x})$)
10: **end for**
11: **return** AL

---

is prefered to sanitize first. It contributes to Max part of algorithm MinMax. The item 'i'selected as Victim item for a particular sensitive itemset X may be present in other sensitive itemset Y but not selected as Victim item for Y due to its higher MCI value than selected Victim item 'i'. Sanitizing X first will reduce the support count of Y also. This is the main idea behind sorting sensitive itemsets. Sensitive transactions are extracted from the original dataset and stored in dataset D' (step 5). The dataset D' transactions sorted according to their relevance value (step 6). Victim items are deleted from selected transactions to sanitize the dataset D' and support count of other affected sensitive and non-sensitive itemsets are updated (step 7–14). $\#IterToSanitize(s_i)$ is the total number of transactions from which Victim item selected for masking of s_i needs to be deleted and $TransToModify$ are those selected transactions. Then sanitized dataset is returned (step 15) after removing the Victim items from selected transactions.

Algorithm 2. MinMax algorithm

Input: S=$\{s_1, s_2, ...s_n\}$, set of sensitive frequent itemsets, NS=$\{ns_1, ns_2, ...ns_m\}$, set of non-sensitive frequent itemsets.
Output: A sanitized dataset.

1: **for** each sensitive itemset $s_i \in S$ **do**
2:     Victim(S_i)← $item_v$ such that $item_v \in s_i$ and $\forall$ $item_k \in S_i$ $MCI(item_v) \leq MCI(item_k)$ {**Min-Part**}
3: **end for**
4: Sort the sensitive itemsets in decreasing order of MCI of their respective victim item {**Max-Part**}
5: D' ←D, where D' made up of transactions containing atleast one of sensitive itemsets.
6: Sort the transactions in D' by their relevance value in descending order
7: **for** each $s_i \in S$ **do**
8:     $\#IterToSanitize(s_i)$= $|T[s_i]|$ - ($| D | \times \delta$)+1
9:     $TransToModify$ ← Select first $\#IterToSanitize(s_i)$ transactions from sorted D' that contains s_i as subset.
10:     **for** each T $\in$ $TransToModify$ **do**
11:         T ← (T - Victim(S_i))
12:         Decrease the support of other affected sensitive and non sensitive itemsets
13:     **end for**
14: **end for**
15: **return** sanitized dataset.

4.1 Time Complexity Analysis

Here, the running time of sanitization process is analysed without considering the running time of the algorithm used to produce frequent patterns i.e. Apriori algorithm [2]. In Algorithm 1, occurence of every frequent 1-itemset in non-sensitive patterns is calculated. Its worst case time complexity is equal to the $O(|I| \times |NS|)$, where $|I|$ is the number of distinct items present in dataset D and $|NS|$ is the number of non-sensitive patterns. In Algorithm 2, process of identification of Victim item for every sensitive itemset takes time $O(|I| \times |S|)$ (step 1–3). Sorting the sensitive itemsets according to the MCI value execute in $O(|V| \times |V|)$ time (step 4). The process of filtering out the sensitive transactions and making the sensitive dataset D' is done in less than $O(|D| \times |S|)$ time (step 5). For sorting the sensitive dataset D' according to the decreasing order of relevance of transactions, transactions are sorted according to increasing order of number of non-sensitive itemsets they contain, which executes in $O(|D'| \times |NS|)$ time (step 6). For each sensitive itemset, the number of iterations to modify the transactions is calculated in $O(1)$ time. To sanitize the dataset, victim items are deleted from the transactions and support count of affected sensitive and non-sensitive patterns are updated which takes $O(|D'| \times |S|)$ time for execution.

4.2 Example

Consider that $S = \{D, CG, CF\}$ is a set of sensitive itemsets randomly selected from frequent itemsets shown in Table 3. Table 2 shows the projected dataset having transactions containing atleast one of these sensitive itemsets. Below are the steps to sanitize the example dataset using proposed algorithm.

1. **Misses Cost impact calculation:**
   $AL = \{MCI(C) = 5, MCI(D) = 3, MCI(F) = 5, MCI(G) = 3\}$.
2. **Victim item selection:**
   $Victim(D) = D$, $Victim(CG) = G$ since $MCI(G)$ is less than $MCI(C)$, $Victim(CF) = C$ since $MCI(C) = MCI(F)$, so any one of C and F can be selected.
3. **Sorting of sensitive itemsets:**
   $S = \{CF, CG, D\}$ is a set of sensitive itemsets sorted in decreasing order of MCI value of their Victim item. Since MCI value of CF is maximum among sensitive itemsets hence selected for masking first.
4. **Sanitization:**
   #IterToSanitize(CF) = $5 - 3 + 1 = 3$ and TransToModify = $\{T2, T3, T4\}$ are selected transactions according to ROT values of transactions.
5. **Support Reduction**
   Deleting item C from these transactions decreases the support count of CG along with complete masking of CF.

Similarly, two other sensitive itemsets are masked. Misses Cost of proposed algorithm MinMax turns out to be 5 with deletion of total 5 items while Greedy, MaxFIA and MinFIA incur Misses Cost 6, 6 and 7 with deletion of 5, 5 and 6 items respectively. It indicates that MinMax preserves better data quality due to less Misses Cost as compared to Greedy, MaxFIA and MinFIA.

Table 4. Characteristics of used datasets

Dataset	Number of transactions	Number of distinct items	Average length of transactions
Chess	3196	76	37.0
Mushroom	8124	120	23.0
BMS-1	59602	497	2.5

5   Experimental Results

Extensive experiments are conducted to evaluate the performance of the proposed algorithm and compared with the existing algorithms namely MaxFIA [16], MinFia [16] and Greedy [5]. All the algorithms are implemented in the JAVA language on Eclipse platform and executed on the Intel®Xeon(R) processor with 64 GB of RAM running Ubuntu 14.04LTS at 2.40 GHz. All of these algorithms completely hide the sensitive itemsets, hence the value of Hiding Failure for all of them is zero. Effect of data sanitization on the dataset's quality is determined by Misses Cost. Three performance parameters are taken into consideration: Misses Cost, Data Loss (in terms of no. of item deleted throughout sanitization process) and Execution Time. Three real-world benchmark datasets Chess, Mushroom and BMS-1 are used in experiments. The Chess and Mushroom datasets are available on Frequent Itemset Mining Dataset Repository present at link

Table 5. Misses Cost with varying percentage of sensitive itemsets

Dataset	MST	Sens_Per	MinMax	Maxfia	Minfia	Greedy
Chess	0.9	1	**155**	203	179	210
		2	**271**	364	281	335
		3	**284**	335	292	313
		4	**223**	279	243	265
		5	**245**	318	252	282
Mushroom	0.4	5	**217**	363	227	252
		6	**221**	310	233	292
		7	**262**	312	269	289
		8	**220**	253	229	237
		9	**281**	305	298	310
Bms-1	0.001	1	**532**	686	623	576
		2	**839**	989	919	885
		3	**1244**	1410	1327	1326
		4	**1393**	1762	1487	1471
		5	**1388**	1560	1441	1490

http://fimi.uantwerpen.be/data/. The another dataset, BMS-1 is click-stream data from a webstore used in KDD-Cup 2000 [10] and accessed from SPMF: An Open-Source Data Mining Library through link http://www.philippe-fournier-viger.com/spmf/index.php?link=datasets.php. The Table 4 shows the characteristics of these datasets. All the experiments are conducted on randomly selected sets of sensitive itemsets. The performance of the proposed algorithm is evaluated by varying minimum support threshold (MST) and percentage of sensitive itemsets (Sens_Per). For each combination of Sens_Per and MST, five samples of sensitive itemsets are randomly drawn. Average value of each performance factor on these five samples is considered for result comparison. In the experiments, the value of MST and Sens_Per parameters are different for each dataset, adjusted based on each dataset's characteristics.

5.1 Varying Percentage of Sensitive Itemsets

The performance of the proposed algorithm is evaluated on the datasets by varying percentage of sensitive itemsets. It is shown in Table 5 that Misses Cost incurred by proposed algorithm on used datasets is less than the other algorithms, hence proposed algorithm, MinMax ensures better quality of data while preserving its privacy. Table 6 shows the number of deleted items by different algorithms to lower the support of sensitive itemsets which concludes that Data Loss by MinMax is less than MinFIA and slightly greater than MaxFIA and Greedy algorithm. Here, Data Loss is measured in terms of item deletion hence it will not result in much higher dropping of data as compared to other algorithms.

Table 6. Number of items deleted with varying percentage of sensitive itemsets

Dataset	MST	Sens_Per	MinMax	Maxfia	Minfia	Greedy
Chess	0.9	1	200	202	205	203
		2	465	434	466	406
		3	514	428	517	399
		4	413	386	402	365
		5	462	431	433	378
Mushroom	0.4	5	5553	4315	5525	4448
		6	5771	6226	5588	5968
		7	6795	5833	6803	5997
		8	5267	5351	5076	5115
		9	7655	6240	7494	6759
Bms-1	0.001	1	2587	2564	2585	2555
		2	4930	4914	4924	4900
		3	7692	7613	7707	7586
		4	8553	7677	8573	8486
		5	10406	10450	10418	10475

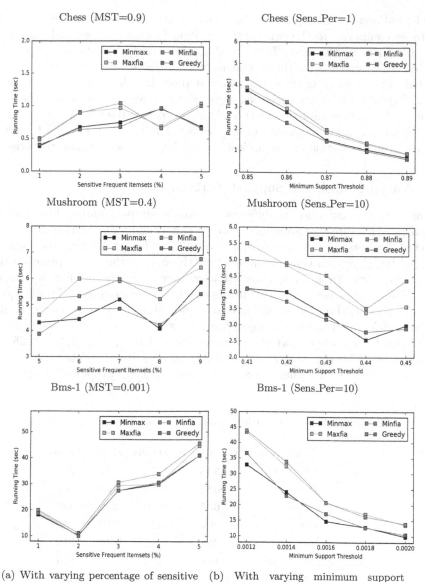

(a) With varying percentage of sensitive itemsets

(b) With varying minimum support threshold value

Fig. 1. Execution time

The loss of non-sensitive patterns due to sanitization is less in the MinMax algorithm as compared to the others. Hence, item loss can be ignored because it does not increase the side effects. It also proves that it is the nature of item deleted which matters for the increment of side effects not the number of items deleted. The number of deleted matters if there is a huge difference between them. Figure 1 (a) shows execution time taken by algorithms on all the three datasets. Execution time of MinMax algorithm is less than MaxFIA and Min-FIA algorithms and commensurate to Greedy algorithm. This is due to sorting of dataset and selection of Victim items according to non-sensitive patterns.

5.2 Varying Minimum Support Threshold

To analyze the influence of different minimum support threshold values on proposed algorithm's performance, further experiments are carried out on the selected datasets. It can be concluded from Table 7 that Misses Cost incurred by MinMax algorithm is less than all other three algorithms that promises better data quality. Table 7 also shows that on increasing the MST value, Misses Cost of algorithms decreases. But this is due to the less number of the frequent itemsets generated on increased MST. It is indicated from Table 8 that number of items deleted by proposed algorithm is less than the MinFIA and greater than MaxFIA and Greedy which concludes that Misses Cost is majorly affected by selection of Victim item & not by total number items deleted during sanitization as discussed earlier. Execution time taken by proposed algorithm is

Table 7. Misses Cost with varying minimum support threshold

Dataset	MST	Sens_Per	MinMax	Maxfia	Minfia	Greedy
Chess	0.85	1	**1545**	1756	2367	1712
	0.86		**1081**	1227	1102	1161
	0.87		**635**	816	680	781
	0.88		**466**	622	560	546
	0.89		**287**	389	314	340
Mushroom	0.41	10	**186**	220	186	203
	0.42		**181**	280	187	187
	0.43		**158**	173	161	165
	0.44		**115**	121	117	122
	0.45		**125**	136	155	131
Bms-1	0.0012	10	**933**	977	962	951
	0.0014		**553**	587	593	572
	0.0016		**348**	395	356	378
	0.0018		**258**	270	270	260
	0.0020		**192**	195	198	206

Table 8. Number of items deleted with varying minimum support threshold

Dataset	MST	Sens_Per	MinMax	Maxfia	Minfia	Greedy
Chess	0.85	1	1029	870	1017	846
	0.86		915	762	931	704
	0.87		564	513	571	515
	0.88		484	436	498	419
	0.89		364	329	370	297
Mushroom	0.41	10	5863	5939	5541	5843
	0.42		6062	5571	6198	5665
	0.43		5952	4927	5815	5014
	0.44		4389	3941	4346	4123
	0.45		5542	3504	5647	3294
Bms-1	0.0012	10	17388	17310	17525	17257
	0.0014		17880	17631	18122	14360
	0.0016		14083	14460	14007	14428
	0.0018		14541	14902	14583	14892
	0.0020		14585	14465	14431	14446

commensurate to Greedy and less than MaxFIA and MinFIA which is shown in Fig. 1 (b) for used datasets. BMS-1 took more time for execution because of its largest dataset size among three.

6   Conclusion

Various heuristic based techniques for sensitive pattern hiding have been proposed by the researchers. In these techniques, maintaining the balance between data quality and data privacy has been the biggest challenge. This paper has proposed a new efficient heuristic technique which preserves better data quality as compared to existing knowledge hiding heuristics. Proposed algorithm considers the impact of Victim item deletion on non-sensitive knowledge while selecting the Victim item and corresponding transaction. This heuristic can conceal all of sensitive itemsets with less Misses Cost as compared to some of existing heuristic based techniques. Experiments show that the proposed technique performs well in terms of execution time on small datasets. It incurs high computational cost for large datasets due to its sequential nature. Future research will intend to improve the proposed algorithm so that data privacy along with good data quality can be achieved on big datasets within real execution time.

References

1. Aggarwal, C.C., Philip, S.Y.: A general survey of privacy-preserving data mining models and algorithms. In: Aggarwal, C.C., Yu, P.S. (eds.) Privacy-Preserving Data Mining. ADBS, vol. 34, pp. 11–52. Springer, Boston (2008). https://doi.org/10.1007/978-0-387-70992-5_2
2. Agrawal, R., Srikant, R., et al.: Fast algorithms for mining association rules. In: Proceedings of the 20th International Conference Very Large Databases VLDB, vol. 1215, pp. 487–499 (1994)
3. Amiri, A.: Dare to share: protecting sensitive knowledge with data sanitization. Decis. Support Syst. **43**(1), 181–191 (2007)
4. Atallah, M., Bertino, E., Elmagarmid, A., Ibrahim, M., Verykios, V.: Disclosure limitation of sensitive rules. In: Proceedings 1999 Workshop on Knowledge and Data Engineering Exchange (KDEX 1999) (Cat. No. PR00453), pp. 45–52. IEEE (1999)
5. Cheng, P., Roddick, J.F., Chu, S.-C., Lin, C.-W.: Privacy preservation through a greedy, distortion-based rule-hiding method. Appl. Intell. **44**(2), 295–306 (2015). https://doi.org/10.1007/s10489-015-0671-0
6. Dai, B.R., Chiang, L.H.: Hiding frequent patterns in the updated database. In: 2010 International Conference on Information Science and Applications, pp. 1–8. IEEE (2010)
7. Gkoulalas-Divanis, A., Verykios, V.S.: An integer programming approach for frequent itemset hiding. In: Proceedings of the 15th ACM International Conference on Information and Knowledge Management, pp. 748–757. ACM (2006)
8. Gkoulalas-Divanis, A., Verykios, V.S.: Exact knowledge hiding through database extension. IEEE Trans. Knowl. Data Eng. **21**(5), 699–713 (2008)
9. Jadav, K.B., Vania, J., Patel, D.: Efficient hiding of sensitive association rules for incremental datasets. Int. J. Innov. Adv. Comput. Sci. (IJIACS) (2014)
10. Kohavi, R., Brodley, C.E., Frasca, B., Mason, L., Zheng, Z.: KDD-cup 2000 organizers' report: peeling the onion. SIGKDD Explor. **2**(2), 86–98 (2000)
11. Lin, C.W., Hong, T.P., Hsu, H.C.: Reducing side effects of hiding sensitive itemsets in privacy preserving data mining. Sci. World J. **2014** (2014)
12. Lin, C.W., Hong, T.P., Yang, K.T., Wang, S.L.: The GA-based algorithms for optimizing hiding sensitive itemsets through transaction deletion. Appl. Intell. **42**(2), 210–230 (2015)
13. Lin, C.W., Zhang, B., Yang, K.T., Hong, T.P.: Efficiently hiding sensitive itemsets with transaction deletion based on genetic algorithms. Sci. World J. **2014**, 13 (2014)
14. Lin, J.C.W., Liu, Q., Fournier-Viger, P., Hong, T.P., Voznak, M., Zhan, J.: A sanitization approach for hiding sensitive itemsets based on particle swarm optimization. Eng. Appl. Artif. Intell. **53**, 1–18 (2016)
15. Moustakides, G.V., Verykios, V.S.: A maxmin approach for hiding frequent itemsets. Data Knowl. Eng. **65**(1), 75–89 (2008)
16. Oliveira, S.R., Zaiane, O.R.: Privacy preserving frequent itemset mining. In: Proceedings of the IEEE International Conference on Privacy, Security and Data Mining, vol. 14, pp. 43–54. Australian Computer Society, Inc. (2002)
17. Oliveira, S.R., Zaïane, O.R.: Protecting sensitive knowledge by data sanitization. In: Third IEEE International Conference on Data Mining, pp. 613–616. IEEE (2003)

18. Öztürk, A.C., Ergenç, B.: Dynamic itemset hiding algorithm for multiple sensitive support thresholds. Int. J. Data Warehous. Min. (IJDWM) **14**(2), 37–59 (2018)
19. Pontikakis, E.D., Theodoridis, Y., Tsitsonis, A.A., Chang, L., Verykios, V.S.: A quantitative and qualitative analysis of blocking in association rule hiding. In: Proceedings of the 2004 ACM Workshop on Privacy in the Electronic Society, pp. 29–30. ACM (2004)
20. Saygin, Y., Verykios, V.S., Clifton, C.: Using unknowns to prevent discovery of association rules. ACM Sigmod Record **30**(4), 45–54 (2001)
21. Saygin, Y., Verykios, V.S., Elmagarmid, A.K.: Privacy preserving association rule mining. In: Proceedings Twelfth International Workshop on Research Issues in Data Engineering: Engineering E-Commerce/E-Business Systems RIDE-2EC 2002, pp. 151–158. IEEE (2002)
22. Sharma, S., Toshniwal, D.: MR-I MaxMin-scalable two-phase border based knowledge hiding technique using MapReduce. Future Gener. Comput. Syst. (2018)
23. Sun, X., Yu, P.S.: Hiding sensitive frequent itemsets by a border-based approach. J. Comput. Sci. Eng. **1**(1), 74–94 (2007)
24. Wang, S.L., Jafari, A.: Using unknowns for hiding sensitive predictive association rules. In: IRI-2005 IEEE International Conference on Information Reuse and Integration 2005, pp. 223–228. IEEE (2005)
25. Wu, J.M.T., Zhan, J., Lin, J.C.W.: Ant colony system sanitization approach to hiding sensitive itemsets. IEEE Access **5**, 10024–10039 (2017)
26. Zamani Boroujeni, F., Hossein Afshari, D.: An efficient rule-hiding method for privacy preserving in transactional databases. J. Comput. Inf. Technol. **25**(4), 279–290 (2017)

Classification and Regression

Classification and Regression

Interpretable Survival Gradient Boosting Models with Bagged Trees Base Learners

Wojciech Jarmulski$^{(\boxtimes)}$ ⓘ and Alicja Wieczorkowska ⓘ

Polish-Japanese Academy of Information Technology, Koszykowa 86,
02-008 Warsaw, Poland
wojciech.jarmulski@pja.edu.pl, alicja@poljap.edu.pl

Abstract. In this paper we present a novel survival analysis modeling approach based on gradient boosting using bagged trees as base learners. The resulting models consist of additive components of single variable models and their pairwise interactions, which makes them visually interpretable. We show that our method produces competitive results often having the predictive power higher than full-complexity models. This is achieved while maintaining full interpretability of the model, which makes our method useful in medical applications.

Keywords: Survival analysis · Gradient boosting · Additive models · Interpretable models

1 Introduction

In survival data we do not know many of the outcome values (e.g. death, graft rejections or disease recurrence in medical studies) because the event might not have occurred within the fixed period of the study or because patients could have become unavailable during the study, i.e. lost to follow-up. In such cases, the date of the last visit (censoring time) provides a lower bound on the survival time. Such datasets are considered censored.

There are many machine learning model adoptions to survival analysis – starting from Cox regression [1], through random survival forests [2] and gradient boosting machines [3], to deep learning models [4]. Cox regression is still one of the most popular methods used in medical literature in survival analysis [5, 6]. Its main strength are interpretability and usefulness in explanatory analysis. One the other hand, it has low predictive power. Machine learning methods help to improve the predictive power of generated models at the cost of losing interpretability. Our aim is to find a method with the best possible predictive power while maintaining full interpretability of the model, which is critical in medical applications.

In this article we present a novel survival analysis modeling approach, which maintains models' interpretability while providing high predictive power comparable to full-complexity models. We achieve that by deriving additive models composed of functions of single predictors and their pairwise interactions as an input. Our models

© Springer Nature Switzerland AG 2020
M. Ceci et al. (Eds.): NFMCP 2019, LNAI 11948, pp. 39–51, 2020.
https://doi.org/10.1007/978-3-030-48861-1_3

are constructed via component-wise gradient boosting where base learners are represented by bagged trees.

The structure of the article is as follows: first we introduce the assumptions of survival analysis and models adoption to this area, then we derive our modelling method fulfilling our criteria. The next part presents the evaluation method of our approach, which is then followed by the results, the presentation and discussion on the properties of our models.

2   Survival Analysis

2.1   Notations

In survival analysis [5], a patient i is represented by a triplet (x_i, δ_i, T_i), where $x_i = (x_{i1}, \ldots, x_{ip})$ is the vector of the patient parameters (characteristics) or the vector of features; T_i indicates time to event of the patient, it is assumed to be non-negative and continuous. If the event of interest is observed, T_i corresponds to the time between the start of the observation and the time of event happening, in this case $\delta_i = 1$, and we have an uncensored observation. If the instance event is not observed and its time to event is greater than the observation time, T_i corresponds to the time between baseline time and end of the observation, and the event indicator is $\delta_i = 0$, and we have a censored observation. Suppose a training set D consists of n triplets (x_i, δ_i, T_i), $i = 1, \ldots, n$. The goal of survival analysis is to estimate the time to the event of interest T for a new patient with feature vector denoted by x by using the training set D.

The survival function $S(t)$ is the probability of a patient surviving longer than t, i.e.:

$$S(t) = P(T > t). \tag{1}$$

The hazard function denoted by $\lambda(t)$ is the instant probability that the event occurs knowing that it did not occur before t. We can define $\lambda(t)$ as:

$$\lambda(t) = \lim_{dt \to 0} \frac{P(t \leq T < t + dt | T > t)}{dt}. \tag{2}$$

The survival function $S(t)$ can be expressed as a function of the hazard at all durations up to t:

$$S(t) = \exp\left(-\int_0^t \lambda(x)dx\right). \tag{3}$$

2.2 Partial Likelihood

A traditional survival model is derived by maximizing the following empirical likelihood $L(f(x_i))$ for right-censoring:

$$L(f(x_i)) = \prod\nolimits_{\forall i, \delta_i = 0} P(T = t_i | f(x_i)) \prod\nolimits_{\forall i, \delta_i = 1} P(T > t_i | f(x_i)), \qquad (4)$$

where $P(T|x)$ is characterized by a parametric distribution and f is derived from the maximization of L. Due to the inclusion of observations whose outcome is unknown (censored), L is referred to as *partial likelihood*. In this work we minimize negative log partial likelihood to adapt a machine learning method to survival analysis, which is the most common approach in literature [7].

2.3 Cox Proportional Hazards Model

Some common approaches attempt to model the hazard function using the proportional hazards assumption. Different functional forms of λ have been considered. Among the most well-known, the semi-parametric Cox proportional hazards model [1] defines λ at time t for an individual with features x_i as:

$$\lambda_i(t|x_i) = \lambda_0(t) \exp(\theta \cdot x_i), \qquad (5)$$

where θ represents parameters.

Under the Cox proportional hazards assumptions, the partial likelihood takes the form:

$$L(\theta) = \prod\nolimits_{j=1}^{n} \left[\frac{\exp(\theta x_j)}{\sum_{i \in R_j} \exp(\theta x_i)} \right]^{\delta_i}, \qquad (6)$$

where R_j is the set of observations at risk at time t_j, i.e. whose event of interest is observed at time t_j or later.

3 Proposed Method

There are numerous machine learning methods whose performance is competitive in comparison to Cox regression [8]. These include random forests, support vector machines, gradient boosting and neural networks, among others. While neural networks and their deep learning variants achieve state-of-the-art results in many applications [9], they are not practical in small data solutions and they are black-box models with a very limited interpretability. Gradient boosting technique is considered one of the most powerful machine learning solutions and its variations are used in many winning solutions in data science competitions [10]. Gradient boosting is an ensemble of weak learners and, as will be shown in the following subsections, imposing some limitations in the construction procedure can lead to fully interpretable models. Naturally, the derivation technique has to be adapted to survival analysis regime.

3.1   Gradient Boosting

Gradient boosting models [11] are built through the search for the optimal prediction function F^* that is defined as:

$$F^* := \mathrm{argmin}_F\, E_{Y,X}[\rho(y, F(x))], \tag{7}$$

where ρ is the loss function differentiable with respect to F. In practice, we usually deal with realizations $(y_i, x_i), i = 1, \ldots, n$; therefore the expectation in (7) is unknown. For that reason gradient boosting algorithms minimize the empirical risk $\mathcal{R} := \sum_{i=1}^{n} \rho(y_i, F(x_i))$ over F instead.

In gradient boosting it is assumed that $F(x)$ follows an additive expansion which takes the form

$$F(x) = \sum_{m=0}^{M} f_m(x) \tag{8}$$

where f is called the *weak* or *base learner*. The following steps are used to minimize R over F:

1. Set an initial estimator $f_0(x)$.
2. For each iteration $m \in \{1, 2, \ldots, M\}$:
   a. Calculate the negative gradient vector g_m of the loss function ρ over F and evaluate it at the estimate of the previous iteration $F_{m-1}(x)$:

$$g_m = (g_{m,i})_{i=1,\ldots,n} := -\left[\frac{\partial \rho(y, F(x_i))}{\partial F(x_i)}\right]_{F(x_i)=F_{m-1}(x_i)}. \tag{9}$$

   b. Calculate estimator $f_m(x)$ by selecting the base learner that best fits the negative gradient vector according to the least squares criterion:

$$f_m(x) = \mathrm{argmin}_f \sum_{i=1}^{n} (g_{m,i}(x) - f(x))^2. \tag{10}$$

   c. Update current estimate by setting:

$$F_m(x) = F_{m-1}(x) + v f_m(x), \tag{11}$$

where $0 < v \leq 1$ is a shrinkage parameter used to control overfitting.

The gradient boosting framework presented above will be used as a base for further adjustments and modifications to derive our target models.

3.2   Additive Representations

Gradient boosting models have the additive form presented in (8). In the standard form each of the base learners f is a function of all input predictor variables $x_1, \ldots, x_p$, which do not provide sufficient interpretability and it is impossible to reliably represent the impact of single predictors on the outcome.

Generalized additive models [12, 13] address this interpretability issue by imposing a limitation that:

$$F(x_1, \ldots, x_p) = f(x_1) + \ldots + f(x_p),$$ (12)

where f_i becomes a function taking a single predictor as an input. In [14] the following extension is proposed which improves models' results while maintaining their interpretability:

$$F(x_1, \ldots, x_p) = f(x_1) + \ldots + f(x_p) + f(x_1, x_2) + \ldots + f(x_{p-1}, x_p).$$ (13)

In this form F is represented by the sum of functions f which model single and pairwise interactions between predictors. To build this form of prediction function F, gradient boosting method presented in Sect. 3.1 needs to be modified: in step 2b we calculate base learners f_m for all single predictors and their pairwise interactions. Then the best fitting base learner is chosen based on the residual sum of squares criterion. This modification version is known as *component-wise gradient boosting* [15].

3.3 Base Learner Function

Base learners are simple regression estimators with a fixed set of input variables and a univariate output. The most common and originally suggested by [11] form of base learners are trees that take all input predictor variables $x_1, \ldots, x_p$. On the other hand, base learners can be as simple as linear models using just one predictor variable as an input.

To maintain interpretability provided by the prediction function in the form (12) and (13), we limit base functions to a function taking only one or two predictors as an input. Following the research in [16], we will use bagged trees which proved to give the best results. Section 5.1 contains further discussion on the choice of bagged trees as base learners.

3.4 Loss Function

Gradient boosting models were adapted for the Cox model by [3]. The loss function is the negative log partial likelihood:

$$\rho(y, F) = - \sum_{i=1}^{n} \delta_i \left[F(x_i) - log \left(\sum_{j:t_j > t_i} e^{F(x_j)} \right) \right].$$ (14)

In our research we use this form of adaptation of gradient boosting to survival analysis because it is most grounded in the literature. There are, however, other adaptation approaches [17].

4 Evaluation

4.1 Datasets

To provide reliable performance results of various models, we evaluated our derived methods on the following real-world datasets:

- STD (Sexually Transmitted Disease) morbidity data [5]. There are 877 patients out of which 60% have a censored outcome.
- RETINOPATHY [18, 19] – a dataset with a trial of laser coagulation as a treatment to delay diabetic retinopathy. Dataset consists of 394 observations with 61% censorship rate.
- METABRIC (Molecular Taxonomy of Breast Cancer International Consortium) [20] is a study that aims to classify breast tumors using molecular signatures in order to find the optimal treatment strategy for patients. The dataset contains clinical information of 1,980 patients and gene expression data. 57.72% of the patients die due to breast cancer over the duration of the study.
- RGBSG – to train our models we use information about 2,232 patients with node-positive breast cancer from Rotterdam and German Breast Cancer Study Groups [21, 22]. 56% of the data is censored.

4.2 Methods

Our evaluation compares our proposed approach against Cox regression and full-complexity tree-based gradient boosting:

- Cox – we treat Cox regression described in Sect. 2.3 as the baseline model,
- GB-SP – our proposed additive model with single predictor functions in the functional form (12), based on bagged trees derived via gradient boosting,
- GB-I – our proposed additive model with single and pairwise interactions between predictors in the functional form (13), modeled with bagged trees and derived via gradient boosting,
- GB-F – full complexity gradient boosting model without any modifications as described in Sect. 3.1.

All gradient boosting methods have been implemented using *mboost* package [23] in R statistical language version 3.6.1.

4.3 Parameters

We use the following choices for parameters' values in models building:

- Number of boosting iterations M is considered to be the most important tuning parameter of boosting algorithms [24]. We determine the optimal value of this parameter by the validation procedure described in detail in Sect. 4.5.
- Shrinkage parameter v – unlike the choice of the optimal iteration, the choice of the shrinkage parameter v has been shown to be of minor importance for the predictive performance of a boosting algorithm [24]. The only requirement is that the value of

υ is small (we set $\upsilon = 0.01$). Small values of υ are necessary to guarantee that the algorithm does not overshoot the minimum of the empirical risk R.

- Bagged trees – for each base learner function we build 100 trees with 3 terminal nodes. We used values recommended by [16] who also show that the outcome result is not sensitive to the variations in these settings.

4.4   Concordance Index

The concordance index or C-index [25] is the standard performance evaluation metric for survival data. It can be regarded as a generalization of the Area Under the Receiver Operating Characteristic Curve (AUROC) that can handle right-censored data and its interpretation is identical to AUROC. C-index is an estimate of the probability that, in a randomly selected pair of comparable patients, the patient whose event occurs first had a worse predicted outcome.

4.5   Validation

For each dataset we set aside 20% of the data as test set to compare results between the models. From the remaining 80% of the training data, 20% was used as a validation set to determine the optimal values of models' parameters described in Sect. 4.3. Parameters which gave the highest value of C-index on validation set were chosen for the final model.

To report the final results of our models' predictions, we bootstrap the test data and calculate the C-index for each of 100 bootstrap samples, which allows us to generate confidence intervals of the results [26].

5   Results and Discussion

Table 1 contains the comparison of the four modeling methods listed in Sect. 4.1 on four real datasets. In line with expectations, Cox regression has the lowest predictive power measured by C-index. Our gradient boosting based model with single predictors (GB-SP) delivered better results than Cox regression in all cases. In two cases (STD, METABRIC) models with interactions (GB-I) had higher predictive power. In the remaining two cases (RETINOPATHY, RGBSG) GB-SP and GB-I had similar predictive properties. We hypothesize that in these cases survival does not depend on any interactions between predictors, and therefore GB-I models do not produce better results. Interestingly, full-complexity gradient boosting models (GB-F) in most cases have lower predictive power than GB-I, with a possible reason that pairwise interactions are enough for these datasets, and also they more precisely capture the relations between predictors. Finally, the results obtained for RGBSG dataset are not substantially different between methods. This is probably due to the fact that linear combinations of predictors are sufficient to model the outcome survival.

Table 1. Predictive power of four modelling methods on four real-time datasets. Results are represented by C-index with 95% bootstrap confidence intervals. Best results are highlighted in bold.

Dataset	Cox	GB-SP	GB-I	GB-F
STD	0.5589 (0.4889, 0.6294)	0.5754 (0.5086, 0.6534)	**0.5936** (0.5196, 0.6899)	0.5745 (0.5098, 0.6386)
RETINOPATHY	0.5704 (0.5020, 0.6415)	**0.6169** (0.5304, 0.6974)	0.6133 (0.5058, 0.7111)	0.5837 (0.4792, 0.6806)
METABRIC	0.5896 (0.4832, 0.6778)	0.6415 (0.5980, 0.6747)	**0.6966** (0.6602, 0.7377)	0.6551 (0.5996, 0.6972)
RGBSG	0.6421 (0.5316, 0.7344)	0.6481 (0.6223, 0.6783)	0.6497 (0.6225, 0.6801)	**0.6548** (0.6252, 0.6826)

Overall results confirm our hypothesis that using GB-I models could be the first method of choice in survival analysis with the caveat that simpler methods like GB-SP or Cox regression can give similar results if there are no complex relations between predictors and the survival rate.

5.1   Base Learners

Single trees are the standard choice as base learners in gradient boosting and could also provide interpretability if we restrict model's functional form to (12) and (13). However, as shown by [14, 16], the usage of bagged trees as base learners leads to better models' performance. Figure 1 presents a qualitative comparison of base learners generated for one sample predictor on METABRIC dataset. Apart from single and bagged trees, it also demonstrates P-splines which are popular in generalized additive models [13]. P-spline base learner has imposed smoothness and is visually most appealing. However, for the extreme values the generated function is raising, which results in improper score values and lowers overall performance. Tree base learners do not have this issue as the generated polylines are flat at the extremes. Additionally, the bagged tree function is visually smoother and closer in shape to the spline, which might explain better performance of these base learners as per [16].

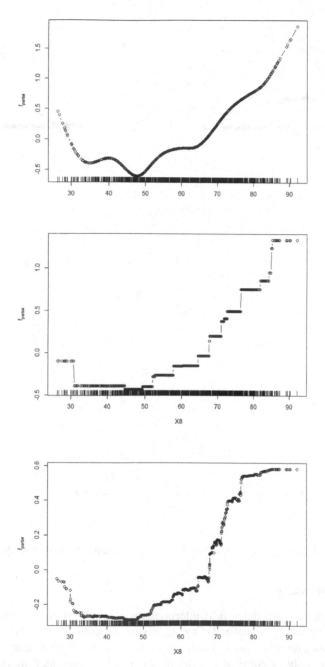

Fig. 1. Visual comparison of base learners – P-spline (top), trees (middle), bagged trees (bottom). Plots present the impact of a sample predictor (x-axis) on the outcome risk score (y-axis).

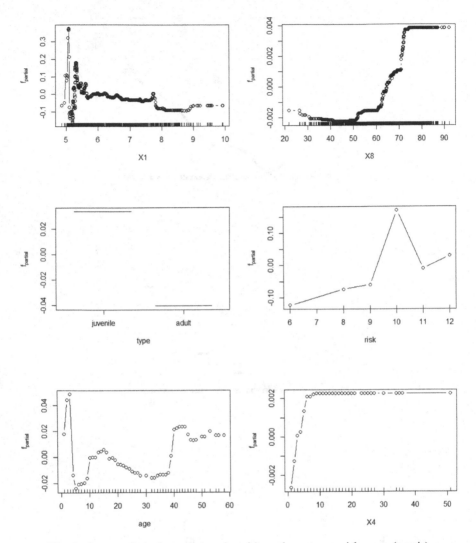

Fig. 2. Impact of single predictors (x-axis) on the outcome risk score (y-axis).

5.2   Interpretability

In this work we treat a model as fully interpretable if the impact of each predictor on the outcome can be clearly visualized. Our additive tree-based gradient boosting models achieve that by producing plots which could be taken for further analysis and interpretation. The additive form of the prediction function presented in (12) and (13) allows for separate graphical interpretation of each single predictor and pairwise interactions between them, and their impact on the outcome risk. Figure 2 shows sample visualizations of the impact of single predictors on the outcome in the form of line charts. Figure 3 visualizes impact of pair of predictors on the outcome risk in the form of heat plots. The outcome risk in our models is directly related to the hazard ratio

(5), and the positive impact of input predictor(s) on it means that patients have higher chance of event occurrence (e.g. death), while negative impact implicates lower probability of the event.

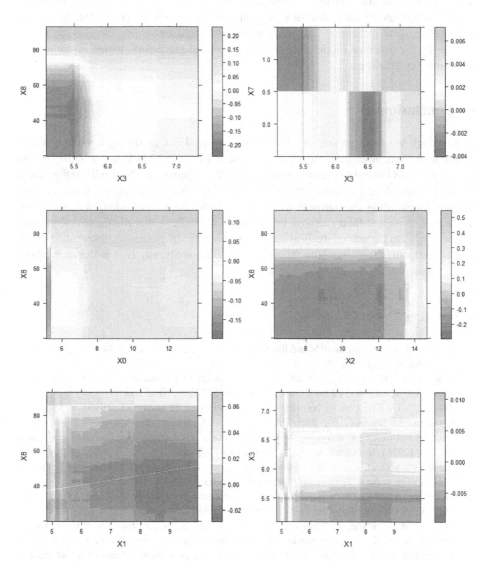

Fig. 3. Heat map plots of pairwise interactions of predictors with their impact on the outcome risk score.

5.3 Feature Selection

Our models creation method has another property – feature selection. In each boosting iteration, the best base learner function is selected, so the outcome prediction function is the sum of only the base learners selected in each boosting iteration. Our base learners are represented by functions taking as an input one or two predictors, therefore, the outcome function uses only predictors which have an impact on the risk score and predictors that do not contribute to the end result are left out from the final model.

The selection of variables is based on the impact on the outcome. However, standard analysis measures, like removal of correlated variables, have to be taken. Otherwise, the outcome model may contain correlated predictors, which might lead to misleading conclusions.

6   Conclusions

In this work we have presented a novel survival analysis modeling approach which allows to achieve high predictive power while maintaining full interpretability of the model, which is critical in medical applications. We achieve that by modifying gradient boosting method and adapting it to survival analysis, using bagged trees as base learners and limiting the outcome functional form to the sum of additive functions of single predictors and their pairwise interactions. We believe that the presented method can be used also outside of medical applications where datasets with observations with missing outcomes (e.g. machine failures) have to be analyzed. This is the area of our future research.

References

1. Cox, D.R.: Regression models and life-tables. J. Roy. Stat. Soc. Ser. B **34**, 187–202 (1972). https://doi.org/10.1111/j.2517-6161.1972.tb00899.x
2. Ishwaran, H., Kogalur, U.B., Blackstone, E.H., Lauer, M.S.: Random survival forests. Ann. Appl. Stat. **2**, 841–860 (2008). https://doi.org/10.1214/08-AOAS169
3. Ridgeway, G.: The state of boosting. Comput. Sci. Stat. **31**, 172–181 (1999)
4. Katzman, J., Shaham, U., Bates, J., Cloninger, A., Jiang, T., Kluger, Y.: DeepSurv: personalized treatment recommender system using a cox proportional hazards deep neural network (2016). https://doi.org/10.1186/s12874-018-0482-1
5. Klein, J.P., Moeschberger, M.L.: Survival Analysis: Techniques for Censored and Truncated Data. Springer, New York (1997). https://doi.org/10.1007/978-1-4757-2728-9
6. Rajkomar, A., Dean, J., Kohane, I.: Machine learning in medicine. N. Engl. J. Med. **380**, 1347–1358 (2019). https://doi.org/10.1056/NEJMra1814259
7. Vock, D.M., Wolfson, J., Bandyopadhyay, S., Adomavicius, G., Johnson, P.E., Vazquez-Benitez, G., et al.: Adapting machine learning techniques to censored time-to-event health record data: a general-purpose approach using inverse probability of censoring weighting. J. Biomed. Inform. **61**, 119–131 (2016). https://doi.org/10.1016/j.jbi.2016.03.009
8. Hothorn, T., Bühlmann, P., Dudoit, S., Molinaro, A., Van Der Laan, M.J.: Survival ensembles. Biostatistics **7**, 355–373 (2006). https://doi.org/10.1093/biostatistics/kxj011
9. LeCun, Y., Bengio, Y., Hinton, G.: Deep learning. Nature **521**, 436–444 (2015). https://doi.org/10.1038/nature14539
10. Chen, T., Guestrin, C.: XGBoost: A Scalable Tree Boosting System (2016). https://doi.org/10.1145/2939672.2939785

11. Friedman, J.H.: Greedy function approximation: a gradient boosting machine. Ann. Stat. **29**, 1189–1232 (2001)
12. Hastie, T.J., Tibshirani, R.J.: Generalized Additive Models, vol. 43. CRC Press, Boca Raton (1990)
13. Wood, S.: Generalized Additive Models: An Introduction with R. CRC Press, Boca Raton (2006)
14. Lou, Y., Caruana, R., Gehrke, J., Hooker, G.: Accurate intelligible models with pairwise interactions. In: Proceedings of the ACM SIGKDD International Conference on Knowledge Discovery and Data Mining; Part F1288, pp. 623–631 (2013). https://doi.org/10.1145/2487575.2487579
15. Buehlmann, P., Hothorn, T.: Boosting algorithms: regularization, prediction and model fitting (with discussion). Stat. Sci. **22**, 477–505 (2007)
16. Lou, Y., Caruana, R., Gehrke, J.: Intelligible models for classification and regression. In: Proceedings of the 18th ACM SIGKDD International Conference on Knowledge Discovery and Data Mining - KDD 2012, p. 150. ACM Press, New York (2012). https://doi.org/10.1145/2339530.2339556
17. Chen, Y., Jia, Z., Mercola, D., Xie, X.: A gradient boosting algorithm for survival analysis via direct optimization of concordance index. Comput. Math. **2013**, 8 (2013)
18. Huster, W.J., Brookmeyer, R., Self, S.G.: Modelling paired survival data with covariates. Biometrics **45**, 145–156 (1989)
19. Blair, A.L., Hadden, D.R., Weaver, J.A., Archer, D.B., Johnston, P.B., Maguire, C.J.: The 5-year prognosis for vision in diabetes. Am. J. Ophthalmol. **81**, 383–396 (1976)
20. Curtis, C., Shah, S.P., Chin, S.-F., Turashvili, G., Rueda, O.M., Dunning, M.J., et al.: The genomic and transcriptomic architecture of 2,000 breast tumours reveals novel subgroups. Nature **486**, 346–352 (2012). https://doi.org/10.1038/nature10983
21. Schumacher, M., Bastert, G., Bojar, H., Hübner, K., Olschewski, M., Sauerbrei, W., et al.: Randomized 2 × 2 trial evaluating hormonal treatment and the duration of chemotherapy in node-positive breast cancer patients. German Breast Cancer Study Group. J. Clin. Oncol. **12**, 2086–2093 (1994). https://doi.org/10.1200/JCO.1994.12.10.2086
22. Foekens, J.A., Peters, H.A., Look, M.P., Portengen, H., Schmitt, M., Kramer, M.D., et al.: The urokinase system of plasminogen activation and prognosis in 2780 breast cancer patients. Cancer Res. **60**, 636–643 (2000)
23. Hothorn, T., Buehlmann, P., Kneib, T., Schmid, M., Hofner, B.: {mboost}: Model-Based Boosting (2018)
24. Hastie, T., Tibshirani, R., Friedman, J.: The Elements of Statistical Learning: Data Mining, Inference, and Prediction, 2nd edn. Springer, New York (2009). https://doi.org/10.1007/978-0-387-84858-7
25. Harrell Jr., F.E., Califf, R.M., Pryor, D.B., Lee, K.L., Rosati, R.A.: Evaluating the yield of medical tests. J. Am. Med. Assoc. **247**, 2543–2546 (1982). https://doi.org/10.1001/jama.1982.03320430047030
26. Efron, B., Tibshirani, R.J.: An Introduction to the Bootstrap. CRC Press, Boca Raton (1994)

Neural Hybrid Recommender: Recommendation Needs Collaboration

Ezgi Yıldırım[(✉)] [iD], Payam Azad, and Şule Gündüz Öğüdücü

Istanbul Technical University, 34467 Sarıyer/Istanbul, Turkey
{yildirimez,sgunduz}@itu.edu.tr, payam.v.azad@gmail.com

Abstract. In recent years, deep learning has gained an indisputable success in computer vision, speech recognition, and natural language processing. After its rising success on these challenging areas, it has been studied on recommender systems as well, but mostly to include content features into traditional methods. In this paper, we introduce a generalized neural network-based recommender framework that is easily extendable by additional networks. This framework named NHR, short for *Neural Hybrid Recommender* allows us to include more elaborate information from the same and different data sources. We have worked on item prediction problems, but the framework can be used for rating prediction problems as well with a single change on the loss function. To evaluate the effect of such a framework, we have tested our approach on benchmark and not yet experimented datasets. The results in these real-world datasets show the superior performance of our approach in comparison with the state-of-the-art methods.

Keywords: Neural networks · Learning latent representation · Recommender systems · Personalization · Hybrid recommenders · Incomplete data

1 Introduction

Online services such as social media and e-commerce have played the key role to derive massive data sources for information systems. Since this information explosion makes users' lives more complicated and even difficult to use such systems, recommender systems aim to offer personalized recommendations to users in order to minimize confusion and increase the chance to reach meaningful information. Based on the available data and the nature of the application domain, there are two main approaches in recommender systems to produce favorable recommendations: collaborative filtering that learn only from past interactions of users and content-based methods that learn the taste of users by using content features. However, both approaches have flaws and favors. While collaborative filtering does not require domain expertise to mine information from data sources and works well for complex objects such as movies, books, music, etc. where variations in taste are much sparse than variations in preferences; content-based

© Springer Nature Switzerland AG 2020
M. Ceci et al. (Eds.): NFMCP 2019, LNAI 11948, pp. 52–66, 2020.
https://doi.org/10.1007/978-3-030-48861-1_4

filtering works better if preference data is sparse and cold-start is an issue. In practice, companies are following a middle way and using hybrid systems of these two approaches. Nevertheless, there are seldom cases of hybrid recommender systems investigated in the literature. Therefore, we present a general framework to use both aspects in a compact deep neural network architecture.

Among the various applied methods, matrix factorization is the most known collaborative filtering approach. Matrix factorization projects user and item into a shared latent space by decomposing the rating matrix into low-dimensional latent factors. To find out an interaction between user and item, the inner product of latent factors are used in recommender systems. In [14], a deep collaborative filtering (DCF) method is proposed to combine probabilistic matrix factorization (PMF) with marginalized denoising auto-encoders (mDA). The latent factors are extracted from the hidden layer of deep networks and they are used to feed matrix factorization components. A collaborative topic modeling approach is proposed by Wang and Blei [18] for recommending scientific articles to online communities. Here, Latent Dirichlet Allocation (LDA) is applied to the user ratings as well as the article contents. Once users and articles are represented as latent factors, matrix factorization is applied to their latent representations to predict user preferences. [12] proposed a context-aware recommendation model, convolutional matrix factorization (ConvMF) that integrates a convolutional neural network (CNN) into PMF. Item representation is obtained from the CNN network that they have trained directly in matrix factorization.

In most of the studies in recommender systems, Deep Neural Networks (DNNs) are used to either get better latent factor representation or integrate auxiliary information into matrix factorization to alleviate the cold-start problem. In contrast to the wide range of researches on the combination of matrix factorization and DNNs, there is relatively little work on employing DNNs to learn the interaction function directly from data. A very first attempt to build a traditional collaborative filtering setup by neural networks [4] simulated matrix factorization by replacing its inner product by a feed-forward neural network, however, it could not be succeeded in benchmark datasets. [9] took this approach one step further because the inner product cannot capture non-linear interactions between users and items. Thus, they proposed a framework named NCF to replace the inner product with non-linear interaction function by a feed-forward neural network and they reported promising results. However, interaction data by itself cannot be sufficient for a challenging recommender system in most cases, auxiliary data is a key factor especially for the systems introducing new users or items at any time. This paper explores the use of DNNs to extract meaningful information from both auxiliary and historical interaction data, then combines them to make better predictions than any single aspects and data sources. Our proposed framework can be extended by not yet experimented auxiliary data and/or by redefining the interaction function using the current data in a flexible manner.

The main contributions of this work are summarized below.

- We devise a general framework for a hybrid recommender system based on DNNs that model latent features of user and item from both auxiliary and interaction data.
- We demonstrate the effectiveness of our NHR approach on the collaboration of self-sufficient recommender models.
- We verify that auxiliary information can significantly improve recommendation quality, especially in large-scale domains. Utilizing auxiliary information can improve not only the success in detecting true interactions but also the ability to correctly rank predictions.
- We show that our NHR approach is essential in the domains that suffer from the severity of cold-starts and rating sparsity due to its stronger contributions to such disadvantaged domains.

Recommendation problems generally suffer from the lack of actual feedbacks given by users a.k.a. explicit feedback. Explicit feedback (via ratings and reviews) is a clear expression of user preferences on items, and it is expressed by direct interactions between system and user. On the other hand, implicit feedback is automatically tracked by the system itself, through inferences about the behavior of the user, such as watching videos, purchasing products and clicking items. Despite the plethora of research over explicit feedbacks; implicit feedbacks are the more realistic case of recommender systems in uttermost situations such as online advertising and online shopping. The reason for the less popularity of using implicit feedbacks is its challenging nature due to the absence of negative interactions. Since we have tested our framework on item prediction problems, we employ negative sampling as discussed in Sect. 3.4 to come through this problem.

2   Neural Hybrid Recommender

In order to build a general framework for both collaborative filtering and auxiliary information, we adopt feed-forward neural networks. Neural networks can model user-item interaction since it has been proven that they are able to learn non-linear relations which is essential for the recommendation of complex objects such as jobs and movies. As suggested in [3], we also utilize wide neural networks for memorization of feature interactions through a wide set of cross-product feature transformations and deep neural networks for better generalization of unseen feature combinations through low-dimensional dense embeddings. Following NCF, we first build a Wide&Deep collaborative filtering approach by combining different neural networks using the same interaction data, then we add auxiliary information by supplementary networks into the system to address the cold-start problem. The names of pure collaborative filtering methods remained as in [9]: GMF (Generalized Matrix Factorization) performing non-linear matrix factorization and MLP (Multi-Layer Perceptron) learning the high-order interaction function. The models trained on auxiliary information are simply named NHR-*type* where *type* refers to the data type that is used for training. We first train multiple self-sufficient neural recommenders independent from each other, then build a framework as an ensemble of all . Even though there is no limitation

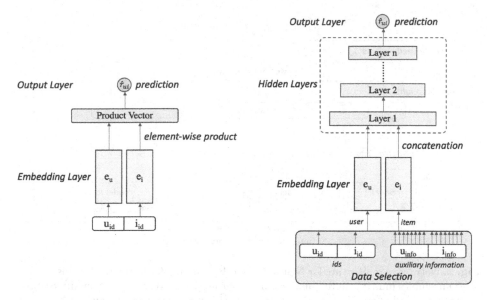

Fig. 1. (left) Representation of neural network realization of matrix factorization; (right) Representation of deep neural recommender networks

on the construction of the models, we can roughly divide what type of networks we use in our experiments into two groups:

- neural network realization of matrix factorization (Fig. 1-left)
- deep neural recommender networks (Fig. 1-right)

Both of the mentioned networks have embedding layers to transform users and items into vector representations. The obtained embedding vectors can be interpreted as the latent vectors of users and items. If we term p_u and q_i as the user latent vector and item latent vector respectively, one can easily define a mapping function as

$$\phi_{mul}\left(p_u, q_i\right) = p_u \odot q_i \tag{1}$$

where ϕ denotes the element-wise product of latent vectors. Then, the next step is to project this product vector to the output layer of the model:

$$\phi_{out}\left(x\right) = \alpha_{out}\left(W^T x + b\right) \tag{2}$$

where $x = \phi_{mul}\left(p_u, q_i\right)$, the output of the multiplication layer in Fig. 1-left, and W, b and α_{out} is the weight vector, bias, and activation function of the layer, respectively. Under the assumptions that the weight vector W is a uniform vector of 1, there is zero bias b in the equation and the activation is an identity function which allows firing the perceptron with the exact value of the input, this project layer acts as a traditional matrix factorization. In order to implement neural network realization of matrix factorization, the weight vector W and the bias b are learned from interactions by the logarithmic loss function in Eq. 5, and

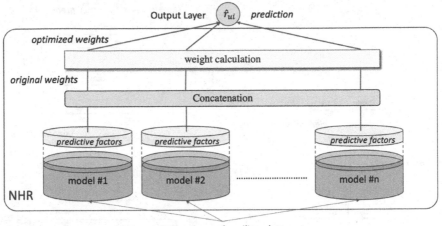

Fig. 2. Network Architecture for Neural Hybrid Recommender Framework. This framework ensembles multiple individual recommender models to reach better quality on predictions. Instead of simply concatenating predictive factors to feed the output layer, we applied a weighting process to define the trade-off between models. One could assign equal importance or learned the importance to the models, but we fine-tuned these weights by optimization on prediction quality on the validation set.

in this way, a non-linear MF approach a.k.a. GMF is obtained. The sigmoid function $\sigma(x) = 1/(1 + e^{-x})$ is used as α_{out} because it restricts each neuron to be in $(0, 1)$ range which meets the expectation for item prediction.

The outputs of the embedding layers on GMF and MLP models are already 1-dimensional vectors because they are fed on inputs of length 1 (*ids* only). However, the embedding layers of deep neural recommender networks trained on auxiliary data (NHR) produce sequences of embeddings w.r.t. sequence length. Average-pooling is a well-known application to gather information exists in the sequence members into a particular form, for example getting sentence embeddings from word embeddings [1, 20], average-pooling is applied to the outputs of embedding layers in these models. Since users and items are represented with several features and every feature has its own embedding space, a concatenation is applied to have one unique latent vector representation for each user-item pair after the average-pooling of embeddings.

Once the latent vectors are obtained for user-item pairs, the following functions are used to generate MLP and NHR models.

$$\phi_1 (\phi_{concat}) = \alpha_1 \left(W_1^T \phi_{concat} + b_1\right)$$
$$\phi_2 (\phi_1) = \alpha_2 \left(W_2^T \phi_1 + b_2\right)$$
$$\vdots$$
$$\phi_n (\phi_{n-1}) = \alpha_n \left(W_n^T \phi_{n-1} + b_n\right)$$

(3)

where α_xs are ReLU activation functions, except the final α_n which is a sigmoid. W_xs are the weight matrices and b_xs are bias vectors as usual.

As reported in [6], the initialization of weights can contribute to convergence and performance of deep learning models. Therefore, we first train all models without prior information till the convergence, then use their parameters to initialize relevant weights on the overall architecture. To combine the models, we simply concatenate the last layers of networks just before the outputs. Since this layer defines the predictive capability of a model, it is generally called as *predictive factors* in literature. We use the original weights of last layers in a weighting process:

$$w \leftarrow \begin{bmatrix} \alpha w^1 & \beta w^2 & ... & \gamma w^n \end{bmatrix} \quad where \ (\alpha + \beta + ... + \gamma) = 1 \tag{4}$$

where w^n denotes the weight vector of nth pre-trained model and $(\alpha, \beta, ..., \gamma)$ is the set of hyper-parameters determining the trade-off between the pre-trained models. The final framework which ensembles multiple self-sufficient neural recommender networks by this weighting process is shown in Fig. 2.

The parameters given in the layer definitions of all models are learned by binary cross entropy loss function given below.

$$\mathcal{L} = - \sum_{(u,i) \in \mathcal{O} \cup \mathcal{O}^-} r_{ui} \log \hat{r}_{ui} + (1 - r_{ui}) \log (1 - \hat{r}_{ui}) \tag{5}$$

where $\mathcal{O}$ denotes the set of observed interactions, and $\mathcal{O}^-$ denotes the set of negative instances. When the loss function is replaced to a weighted squared loss, the proposed framework can be easily applied to explicit datasets as well.

3 Experiments

3.1 Datasets

To conduct our experiments, we worked on two real-world problems: movie recommendation and job recommendation. For the movie recommendation task, we applied our approach to a benchmark movie rating dataset enriched by movie subtitles. The statistics of the experimented datasets are listed in Table 1.

MovieLens and OPUS. MovieLens [8] includes 5-star ratings of movies and some categorical properties of users and movies. It contains $1M$ ratings, $3.8K$ movies and $6K$ users in total. Users have at least 20 ratings. 5-star explicit ratings are converted to implicit feedback by treating a rating is the indicator of user-item interaction, so all ratings in the dataset are considered to be 1. OPUS subtitles dataset [15] describes a collection of translated movie subtitles from http://www.opensubtitles.org/. It composes of bitexts from many language pairs. English subtitles are used to supply more convenient contents for movies. 2581 movies out of 3706 (69.64%) in the rating dataset have subtitles. The movie subtitles in the OPUS dataset are utilized for item representation while the categorical properties of user profiles in the MovieLens for user representation. For more information about the categorical features, we advise readers to visit https://grouplens.org/datasets/movielens/1m/.

Table 1. Statistics of the experimented datasets

Dataset	Type	Interaction	Item	User	Sparsity
MovieLens	Movie	1,000,209	3,706	6,040	95.53%
Kariyer	Job	383,434	16,134	20,283	99.88%

Kariyer. This online recruiting dataset used in our study is obtained through a collaboration with Kariyer.Net, the largest online recruiting website in Turkey. Candidates use *Kariyer.Net* to find a suitable job, and recruiters use *Kariyer.Net* to find the right candidate for a job on behalf of their companies. We evaluate our proposed model on this dataset constructed of user profiles, job postings and user behaviours on this website in a limited time frame, a one-week period. Each user has at least 20 applications. It consists of $383K$ applications, $20K$ candidates for $16K$ jobs in total. The application history of users is used as the interaction data in job recommendation, and the properties of jobs and candidates as the auxiliary data.

The dataset consists of numerical and categorical data types besides text data (see Sect. 3.2 for further details). To make its numerical and categorical features meaningful and applicable for the recommendation task, we first applied some preprocessing steps such as normalization and noise cleaning, we then performed feature transformations. Features exist in the dataset are listed in Table 2 along with their types and representations as a reference for rest of the paper. For normalization of numerical values that are measured on different scales, outliers are removed by considering their statistics, then they are adjusted to the $[0, 1]$ scale.

Table 2. Features in the Kariyer.Net dataset

Feature (candidate)	Feature (job)	Feature type	Representation
Military service Work status Gender Driving licence	Gender Hidden posting Position type Language Driving licence	Categorical	One-hot encoding
Education Faculty University Province	Education Military service Industries Provinces	Categorical	Multi-hot encoding
Age	Min experience Max experience Position level Hiring capacity	numerical	$\{x \mid 0 \leq x \leq 1\}$
Experiment position	Name Qualification Explanation	Text	Sequence of indexes in a fixed-size hashing space

Neural networks, as well as many other machine learning algorithms, require numeric input and output variables. To that end, the most primitive solution to use categorical features is to transform them into integer labels, a.k.a. integer encoding, where each category is represented by unique numbers. However, integer values have ordinal relationships between each other, whereas no such relation exists in categorical variables, so the learning process may result in poor performance. Therefore, we converted categorical features into one-hot or multi-hot representations that work better with learning algorithms. The occurrence statistics per multi-hot feature are given in Table 3.

Table 3. Occurrence statistics of features for each multi-hot feature in the Kariyer.Net dataset

Object	Feature	Max	Mean	Std
Jobs	Education	12	4.1100	2.2337
	Provinces	82	2.5023	8.5675
	Industries	6	1.4352	0.9516
	Military service	4	1.4525	0.6708
Candidates	Education	2	1.0381	0.1913
	Faculty	7	1.3930	0.6402
	University	6	1.3644	0.5945
	Province	5	1.0390	0.2041

3.2   Handling Text Data

To make the text data suitable to feed neural networks, we need to convert raw texts into numeric vectors. In the simplest approach, using a simple dictionary for this purpose could lead to extremely sparse representations due to the huge size of vocabulary. Thus, we exploited the advantage of a hash function which converts a raw text to a sequence of indexes in a fixed-size hashing space. Note that some words may be assigned to the same index according to the hash function. The dimension of hashing space is in relation to the overlapping rate of distinct words and the dimension of embedding layers. By considering the pros and cons, we set the dimension of hashing space to $1K$ in the experiments after evaluating its effect on overall performance and complexity.

Since the inputs to the neural networks have to be in the same size for all iterations, we examined the mean (μ) and the standard deviation (σ) of sequence lengths of text features. Then, the feature-specific input lengths are defined as $\mu + \sigma$ for each text feature in the datasets.

3.3   Evaluation Process

In order to split the dataset into the train and test sets, we preferred *leave-one-out* evaluation which has been widely applied in many works [2,9,10,16], especially where sparse datasets are subjected. The latest interaction of each user is held-out to compose a test set, while the remaining interactions are used for training. The last interaction of each user in the train set is used for hyper-parameters tuning.

Since ranking every user-item pair amongst the test pairs are very time-consuming and not possible to run in real-time. Therefore, as in similar studies [5,9,13] we randomly sampled 100 items per user and rank them by probability of interaction. To measure the quality of ranking, we used well-known evaluation metrics: Hit Ratio (HR) and Normalized Discounted Cumulative Gain (NDCG). We applied both metrics on a truncated list including top-10 ranked test items for each user. Due to the fact that the users have one interaction in the test set, HR@k is simplified in our experiments as follows:

$$HR@k = \begin{cases} 1/k, & \text{if } r_{test}(u,i) \in R_k \\ 0, & \text{otherwise} \end{cases} \tag{6}$$

where $r_{test}(u,i)$ and R_k define the interaction with the item i and the list of top-k recommended items for the user u. In addition to HR@k, NDCG@k is reinterpreted as well in our experiments because ideal discounted cumulative gain $(IDCG_k)$ in position k is equal to 1 in our evaluation setup. Therefore, NDCG@k is redefined as:

$$NDCG@k = \frac{DCG@k}{IDCG@k} = \sum_{i=1}^{k} \frac{2^{r(u,i)} - 1}{\log(i+1)} \tag{7}$$

where $r(u,i)$ is 1 if the user u interacted with the ith item of the top-k list and 0 otherwise. The results are reported by the mean of user scores.

HR gives a shallow understanding of success by considering if the interacted item is in the top-10 list or not whereas NDCG helps for a better understanding by setting higher scores to hits at higher ranks.

3.4 Negative Sampling

In most of the cases, implicit feedback refers to positive inference of user interaction or user interest. To handle the absence of negative feedback, many studies have either assumed all unobserved cases as negative feedback or sampled negative instances from them. In this work, we also apply the latter approach to generate a set of negative feedback by sampling four negative instances per positive instance. Unlike the evaluation process, we randomly sampled negative training instances in real-time, just before each epoch starts. This allows our system to learn as much as possible from different instances and increases the utility of dataset without interfering with its feasibility.

3.5 Baselines

We compared our proposed approach NHR to the following methods:

- **PopRank** is a non-personalized popularity based recommendation method. Items are ranked by their popularity which is determined by the number of interactions and recommended to all users with the same order.
- **BPR** [16] is a highly competitive pairwise ranking method which works well for implicit feedbacks. It optimizes the matrix factorization model with a pairwise ranking loss.
- **ALS** [11] is also a matrix factorization algorithm for item recommendation. It works in parallel and effective for large-scale collaborative filtering problems which suffer from the sparseness of the rating data.
- **GMF** [9] is a neural network realization of matrix factorization. Besides being a part of NCF, it can be employed as a complete recommender system.
- **MLP** [9] is also a part of NCF that learns user-item interaction function by neural networks, Like GMF, it is a standalone recommender system.
- **NCF** [9] is a state-of-the-art neural network based collaborative filtering method which combines GMF and MLP methods. No matter that has very promising results for item prediction, it is a pure collaborative filtering method which benefits from only interaction data and does not regard cold-starts that is a very common case for real-world recommendation tasks.

3.6   Parameter Setting

We implemented our proposed framework using PyTorch. All individual models had been learned by optimizing the logarithmic loss of Eq. 5 because we tested them on an item prediction setup. To determine the hyper-parameters of the methods, we conducted intensive tests on validation data.

For individual models that are trained without any prior information, we set model parameters with a Xavier initialization , then optimize them with Adam optimizer which employs an adaptive learning rate for faster convergence. The learning rate is set to 0.001 and the momentum for Adam optimizer to 0.9 which is the default setting.

We tested a bunch of different batch size but found the 128 is the best performing setup for all, except the model trained on text data. Because the embedding size for the text data is quite large and hard to fit on even comparatively large computer memories, we adopt the batch size of 32 for them.

We evaluated the predictive factors of $\{8, 16, 32, 64\}$. We employed three hidden layers for interaction-specific networks, for example, if the number of predictive factors is set to 8, then the size of hidden layers are selected in the order of $32 \to 16 \to 8$ from the top on down and the embedding size is 16 in this setup, as a matter of course. For the networks trained on auxiliary data, we used two hidden layers and intuitively set embedding size to be 128 for movie subtitles, 4 for job titles and candidate past-positions, and 16 for job qualifications, job explanations and candidate experiments. To treat equally, we set the α parameter of NCF which defines the trade-off between GMF and MLP by optimization as we did for our NHR methods.

Table 4. Performance of HR@10 and NDCG@10 w.r.t. the number of predictive factors (*pf*) on different datasets. Here are the abbreviations used to shrink the result table due to the limited space; *ds*: Dataset, *ML*: MovieLens, *Ka*: Kariyer, *mt*: Metric, *PR*: PopRank, *cat.*: categorical, *comb.*: combined, and *Im.*: Improvements

ds	pf	mt	Baselines						NHR			Im.%
			PR	BPR	ALS	GMF	MLP	NCF	cat.	text	comb.	
ML	8	HR	0.4512	0.5331	0.6076	0.6247	0.6522	0.6560	–	–	**0.6718**	2.41%
		NDCG	0.2546	0.3027	0.3488	0.3528	0.3789	0.3807	–	–	**0.3943**	3.57%
	16	HR	0.4512	0.5886	0.6545	0.6714	0.6626	0.6828	–	–	**0.6946**	1.73%
		NDCG	0.2546	0.3426	0.3886	0.3945	0.3890	0.4057	–	–	**0.4126**	1.7%
	32	HR	0.4512	0.6040	0.6826	0.6757	0.6728	0.6874	–	–	**0.6979**	1.53%
		NDCG	0.2546	0.3564	0.4150	0.3936	0.3986	0.4053	–	–	**0.4147**	2.32%
	64	HR	0.4512	0.6108	0.6912	0.6763	0.5190	0.6798	–	–	**0.6964**	2.44%
		NDCG	0.2546	0.3621	0.4290	0.4052	0.2857	0.4077	–	–	**0.4176**	2.43%
Ka	8	HR	0.3231	0.7399	0.5137	0.8249	0.7448	0.8594	0.8821	0.8624	**0.8834**	2.79%
		NDCG	0.1875	0.5067	0.3237	0.5719	0.5592	0.6204	**0.6368**	0.6188	0.6354	2.64%
	16	HR	0.3231	0.7874	0.6166	0.8357	0.8021	0.8695	0.8890	0.8730	**0.8917**	2.55%
		NDCG	0.1875	0.5560	0.4034	0.6041	0.5564	0.6402	0.6571	0.6426	**0.6579**	2.76%
	32	HR	0.3231	0.7934	0.7013	0.8121	0.8100	0.8658	0.8851	0.8703	**0.8875**	2.51%
		NDCG	0.1875	0.5629	0.4740	0.5870	0.5471	0.6369	0.6537	0.6411	**0.6562**	3.03%
	64	HR	0.3231	0.7922	0.7627	0.7841	0.8205	0.8621	0.8800	0.8678	**0.8841**	2.55%
		NDCG	0.1875	0.5608	0.5394	0.5624	0.5519	0.6334	0.6505	0.6378	**0.6536**	3.19%

3.7 Performance Results

In our NHR experiments, we group auxiliary information sources into three categories: categorical, text, and a combination of them. Kariyer dataset includes many data types: free-text, numerical values, binary, single-label, and multi-label categorical features. In order to handle all different types during the learning process, we first apply general pre-processing steps such as outlier removal, tokenization, etc. We then normalize numerical values and transform binary and categorical features into one-hot and multi-hot representations. All these features are considered *categorical* for simplicity. We also convert raw text features to hash vectors which refer to *text* data source as explained in Sect. 3.2; Both networks trained on the categorical and text data sources are first incorporated into NCF alone (NHR-categorical and NHR-text respectively), then together to embody the most extensive NHR model (NHR-combined). As for Movie-Lens dataset, users are represented with categorical features whereas movies are represented with text features. This results in having one auxiliary network (NHR-combined) which combine the categorical and the text data sources at the same time. Thus, we could report one experiment on NHR for the movie recommendation task.

Table 4 shows the recommendation performance of the compared methods with respect to the number of predictive factors. The results are given in HR@10

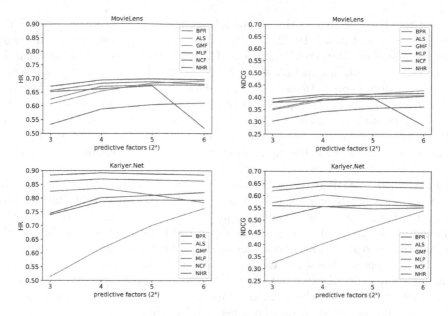

Fig. 3. The effect of number of predictive factors on recommendation performance

and NDCG@10. BPR and ALS methods have the same latent factor size as
the predictive factors in neural network models. By doing so, we use the same
predictive capability for all baselines except PopRank to make a fair comparison between them. Figure 3 shows the HR and NDCG changes per evaluated
methods by changing the number of predictive factors (a.k.a. latent factors in
ALS and BPR). As expected increasing predictive capability of models generally increases the performance of methods until they have reached the limit of
falling into over-fitting. The size of predictive factors may drastically affect the
complexity of the MLP model since the size of all preceding layers is determined
by the predictive factors given in the initialization of models as explained in
Sect. 3.6. This is visible on the severe degradation of both HR and NDCG scores
on MovieLens dataset when predictive factors are set to 64 (2^4)). The methods
including MLP along with other models have the advantage of generalization
with the contribution of their other components, so such models as NCF and
NHR did not affected by this deterioration. Since the job recommendation problem is far more complex than movie recommendation, such complex models are
less prone to over-fitting on this domain as observed in Fig. 3. Regarding the
comparison of methods with the same predictive factor size, PopRank has the
weakest performance amongst the other methods. It is already expected because
it is incapable to make personalized suggestions. Since 0.001-level improvements
are already found to be significant for similar tasks such as click-through rate
(CTR) prediction [3,7,17,19], one can easily say that NHR is significantly out-
performing the state-of-the-art matrix factorization methods, ALS and BPR, by
a large margin in both metrics, and it is also consistently superior to the most

competitive baseline NCF. NHR on MovieLens and Kariyer achieved 2.03% HR-2.51% NDCG and 2.60% HR-2.91% NDCG relative improvements on average over their NCF counterparts, relatively. NHR gains more generalization capability through merging interaction and auxiliary data. In addition to more accurate hits on top-10 predictions, the results show that NHR systems could better learn to rank items in the top-10 lists by uprising the test interaction amongst the other predictions since NDCG scores are improved by larger steps. The NHR-combined results on job recommendation clearly shows that adding new auxiliary data even with the same learning function can enhance the overall recommendation performance.

Even though NHR-text system improves the recommendation quality, it underperforms NHR-categorical because of its model complexity. Besides the inevitable large size of the embedding layer, the Kariyer dataset is extremely sparse and interaction data is not enough to feed such a network in fact. With more data, we expect to have more contribution from text data.

The last but not the least, the results are more promising for the job recommendation. Since Kariyer dataset suffers from a severe sparsity and a high frequency of cold-starts, the auxiliary data and the cooperation of models can fill in this information shortage about user preferences.

4   Conclusion

In this work, we explored DNNs for hybrid recommender systems. We devised a general framework NHR that model user-item interactions by combining auxiliary and historical data. We showed that every variation of NHR outperforms state-of-the-art collaborative filtering methods as expected, but NHR also gives us the chance to alleviate deficiencies to be dependent on single aspects or data sources. It does not require to train complete architecture from scratch. Instead, it allows self-sufficient recommender models to speak for themselves by a weighting process which learns the capabilities of its components.

In the next phase of the study, we would like to test our approach on explicit datasets and use pre-trained vector space models such as document vectors for text features since learning of embedding layers directly effects the model complexity and training time. Since average pooling leads to the loss of sequential property of natural language texts, we would like to improve our text models by using more elaborated architectures such as LSTMs and CNNs to exploit sequence information and interrelation of words.

Acknowledgements. This study is part of the research project supported by the Scientific and Technological Research Council of Turkey (TÜBİTAK) (Project No: 5170032). This work was also supported by the Research Fund of the Istanbul Technical University (Project Number: BAP-40737). We would like to thank Kariyer.Net for providing us with the online recruiting dataset used in the paper.

References

1. Adi, Y., Kermany, E., Belinkov, Y., Lavi, O., Goldberg, Y.: Fine-grained analysis of sentence embeddings using auxiliary prediction tasks. arXiv preprint arXiv:1608.04207 (2016)
2. Bayer, I., He, X., Kanagal, B., Rendle, S.: A generic coordinate descent framework for learning from implicit feedback. In: Proceedings of the 26th International Conference on World Wide Web, pp. 1341–1350. International World Wide Web Conferences Steering Committee (2017)
3. Cheng, H.T., et al.: Wide & deep learning for recommender systems. In: Proceedings of the 1st Workshop on Deep Learning for Recommender Systems, pp. 7–10. ACM (2016)
4. Dziugaite, G.K., Roy, D.M.: Neural network matrix factorization. arXiv preprint arXiv:1511.06443 (2015)
5. Elkahky, A.M., Song, Y., He, X.: A multi-view deep learning approach for cross domain user modeling in recommendation systems. In: Proceedings of the 24th International Conference on World Wide Web, pp. 278–288. International World Wide Web Conferences Steering Committee (2015)
6. Erhan, D., Bengio, Y., Courville, A., Manzagol, P.A., Vincent, P., Bengio, S.: Why does unsupervised pre-training help deep learning? J. Mach. Learn. Res. **11**(Feb), 625–660 (2010)
7. Guo, H., Tang, R., Ye, Y., Li, Z., He, X.: DeepFM: a factorization-machine based neural network for CTR prediction. arXiv preprint arXiv:1703.04247 (2017)
8. Harper, F.M., Konstan, J.A.: The movielens datasets: history and context. ACM Trans. Interact. Intell. Syst. (TiiS) **5**(4), 19 (2016)
9. He, X., Liao, L., Zhang, H., Nie, L., Hu, X., Chua, T.S.: Neural collaborative filtering. In: Proceedings of the 26th International Conference on World Wide Web, pp. 173–182. International World Wide Web Conferences Steering Committee (2017)
10. He, X., Zhang, H., Kan, M.Y., Chua, T.S.: Fast matrix factorization for online recommendation with implicit feedback. In: Proceedings of the 39th International ACM SIGIR Conference on Research and Development in Information Retrieval, pp. 549–558. ACM (2016)
11. Hu, Y., Koren, Y., Volinsky, C.: Collaborative filtering for implicit feedback datasets. In: 2008 Eighth IEEE International Conference on Data Mining, ICDM 2008, pp. 263–272. IEEE (2008)
12. Kim, D., Park, C., Oh, J., Lee, S., Yu, H.: Convolutional matrix factorization for document context-aware recommendation. In: Proceedings of the 10th ACM Conference on Recommender Systems, pp. 233–240. ACM (2016)
13. Koren, Y.: Factorization meets the neighborhood: a multifaceted collaborative filtering model. In: Proceedings of the 14th ACM SIGKDD International Conference on Knowledge Discovery and Data Mining, pp. 426–434. ACM (2008)
14. Li, S., Kawale, J., Fu, Y.: Deep collaborative filtering via marginalized denoising auto-encoder. In: Proceedings of the 24th ACM International on Conference on Information and Knowledge Management, pp. 811–820. ACM (2015)
15. Lison, P., Tiedemann, J.: Opensubtitles 2016: extracting large parallel corpora from movie and TV subtitles. In: Proceedings of the 10th International Conference on Language Resources and Evaluation (2016)
16. Rendle, S., Freudenthaler, C., Gantner, Z., Schmidt-Thieme, L.: BPR: Bayesian personalized ranking from implicit feedback. In: Proceedings of the Twenty-Fifth

Conference on Uncertainty in Artificial Intelligence, pp. 452–461. AUAI Press (2009)

17. Song, W., et al.: Autoint: automatic feature interaction learning via self-attentive neural networks. arXiv preprint arXiv:1810.11921 (2018)
18. Wang, C., Blei, D.M.: Collaborative topic modeling for recommending scientific articles. In: Proceedings of the 17th ACM SIGKDD International Conference on Knowledge Discovery and Data Mining, pp. 448–456. ACM (2011)
19. Wang, R., Fu, B., Fu, G., Wang, M.: Deep & cross network for ad click predictions. In: Proceedings of the ADKDD 2017, p. 12. ACM (2017)
20. Wieting, J., Bansal, M., Gimpel, K., Livescu, K.: Towards universal paraphrastic sentence embeddings. arXiv preprint arXiv:1511.08198 (2015)

Discovering Discriminative Nodes for Classification with Deep Graph Convolutional Methods

Liana-Daniela Palcu[1](✉), Marius Supuran[1](✉), Camelia Lemnaru[1](✉),
Mihaela Dinsoreanu[1](✉), Rodica Potolea[1](✉), and Raul Cristian Muresan[2](✉)

[1] Computer Science Department, Technical University of Cluj Napoca,
Cluj-Napoca, Romania
ldpalcu@gmail.com, marius.supuran@yahoo.com, camelia.lemnaru@cs.utcluj.ro,
mihaela.dinsoreanu@cs.utcluj.ro, rodica.potolea@cs.utcluj.ro
[2] Transylvanian Institute of Neuroscience, Cluj-Napoca, Romania
raul.muresan@gmail.com

Abstract. The interpretability of Graph Convolutional Neural Networks is significantly more challenging than for image based convolutional networks, because graphs do not exhibit clear spatial relations between their nodes (like images do). In this paper we propose an approach for estimating the discriminative power of graph nodes from the model learned by a deep graph convolutional method. To do this, we adapt the Grad-CAM algorithm by replacing the part which heavily relies on the 2D spatial relation of pixels in an image, with an estimate of the node importance by its appearance count in the result of the Grad-CAM. Our strategy was initially defined for a real-world problem with relevant domain-specific assumptions; thus, we additionally propose a methodology for systematically generating artificial data, with similar properties as the real-world data, to assess the generality of the learning process and interpretation method. The results obtained on the artificial data suggest that the proposed method is able to identify informative nodes for classification from the deep convolutional models.

1 Context and Motivation

Model interpretability can be important for several reasons: first, it builds trust and confidence in machine learning models when applied to sensible problems (e.g. medical diagnosis and prognosis, terrorism prediction, credit assessment, etc). In such domains, if the model can explain its decisions, it is easier to asses its fairness (does not discriminate against protected groups), privacy-compliance, robustness and the ability to identify causality [1]. Second, it is a potentially powerful tool for generating new domain knowledge in "difficult" domains, such as neuroscience. For example, interpretable models could provide new insights into understanding the effect of alcohol on the brain. In the same line of argument, if the performance of the model beats human performance (e.g. chess, Alpha-GO), machine-driven instruction could be used to help humans improve their

M. Ceci et al. (Eds.): NFMCP 2019, LNAI 11948, pp. 67–82, 2020.
https://doi.org/10.1007/978-3-030-48861-1_5

skills. Last, but not least, interpretability can be thought of as a useful tool for understanding and correcting model errors.

In general, we are faced with a trade-off between performance and interpretability. Graph classification is normally a domain which requires the application of complex learning models, such as deep neural networks, which are not interpretable by nature. Several relevant attempts have been made to interpret complex models post-hoc (briefly reviewed in Sect. 2). However, most approaches focus on tabular inputs, or inputs with a known, structured or hierarchical relation between the elements (e.g. the 2D spatial relation between pixels in an image or the 1D temporal relation between words in a sentence). For graph data, we do not have such spatial or temporal semantics to work with, which makes interpreting any model built on such data even more difficult.

The starting point of our research is rooted in neuroscience: trying to identify neurons in the brain which are most affected by alcohol consumption, and which separate between non-alcohol and alcohol affected brain states. Graph/network analysis methods that are applied to this problem need to produce interpretable models, because their aim is to help understand brain behavior. The starting hypothesis is that there is a small subset of neurons whose connection weights are affected by alcohol, and those neurons are responsible for changing the overall behavior and response to alcohol. While initially driven exclusively by this hypothesis, we assess that the methods investigated are applicable to any graph classification problem. We propose a method for graph data classification and model interpretation which generates class-specific relevance heatmaps for the nodes in the graph by applying a modified version of Grad-CAM [2] – an interpretability method initially designed for image CNNs.

The rest of the paper is organized as follows: Sect. 2 overviews the relevant interpretation strategies from literature. Section 3 presents the proposed method, which is evaluated in Sect. 4. The last section contains concluding remarks.

2   Related Work

Some classification models (e.g. decision trees, logistic regression) are inherently interpretable. For the others, which have a black box behavior, interpretability methods can be divided into model-agnostic and model-specific [3]. The first category encompasses methods which can be applied to any classification model, and generally focus either on explaining a model by computing feature relevance scores – globally [4–6], or at instance level [7,8] – or try to build a global or a local interpretable surrogate model, such as LIME [9].

In the context of Convolutional Neural Network (CNN) models, agnostic interpretability methods do not exploit that such models learn new features and concepts in their hidden layers and are computationally inefficient, because they do not use gradient values [3]. For interpreting CNN models, recent works in literature focus either on *perturbing the input (or hidden network layers) and observing the corresponding changes* – generally computationally intensive and can show instability to surprise artifacts (a line of research closely related to adversarial attacks on CNN architectures) – or *leveraging gradient values to infer feature importance* – computationally efficient, but poses challenges when propagating gradients back through non-linear and re-normalization layers.

[10] proposes the use of deconvolution to identify which part of an image most strongly activates a specific unit in the network: typically, all neurons except one are set to zero in the high level feature map corresponding to the layer of that unit, and we perform a backward pass through the CNN down to the input layer. The resulting reconstructed image shows which part of the input image most strongly activates that unit. Class specific saliency maps [11] are generated by computing the gradient of the class score with respect to the input image. The intuition is to use the gradients to identify input regions that cause the most change in the output. The main difference between the last two techniques is how gradients are passed through non-linear layers such as the Rectified Linear Unit (ReLU): in [11] the gradients of neurons with *negative input* are suppressed, while in [10] the gradients of neurons with incoming *negative gradients* are suppressed. Guided backpropagation [12] combines both strategies, by suppressing the flow of gradients of both negative input and negative gradient neurons.

Class Activation Maps (CAMs) [13] identify the image regions used by a CNN to discriminate between different categories. It can only be applied to a limited set of CNNs and it alters the architecture by adding at the end a Global Average Pooling layer (GAP) and then a fully-connected layer. This is done to preserve the localization ability of any network, which is lost using fully connected layers. However, this change could affect the performance of the model.

A significant shortcoming of the methods presented above is that they do not address re-normalization layers, such as max-pooling. Propagating gradients back through such a layer can be problematic since the functions used are not generally differentiable. Grad-CAM [2] tries to circumvent this problem by relying on the activation maps of the final convolutional layer to build a down-sampled relevance map (heatmap) of the input pixels, which is then upsampled to obtain a coarse relevance heatmap of the input.

3   Interpreting Graph Convolutional Network Models with Grad-CAM

In the up-sampling step, Grad-CAM performs a bi-linear interpolation between neighboring pixels, which is computationally efficient and produces good results for images, but cannot be directly applied to graphs. Consequently, we modify Grad-CAM to address this and allow the generation of class-relevant heatmaps containing estimates of the each node's importance to a specific class. We integrate our solution with the Deep Graph Convolutional Neural Network model (DGCNN) [14]. As the ultimate goal of the strategy is the identification of the relevant nodes in the classification decision, we propose a preprocessing step which consists of removing potentially non-informative edges.

3.1 Graph Sparsification

Sparsification is motivated by the assumption that not all edges are informative, and that small weight edges represent noise. Consequently, sparsification eliminates a certain amount of small weight edges, with the hope of improving classification accuracy and model interpretability. Let $G(V, E)$ be a complete weighted graph, where V represents the set of nodes and E represents the set of edges, each edge being given by $e_i(u, v, w_i)$, with $u, v \in V$ and $w_i \in \mathbb{R}^+$. Let Sum_G be the sum of all the weights from the graph. We sort the edges in descending order by weight, $<e_1, e_2, ..., e_n>$, where $w_1 > w_2 > ... > w_n$, and considering this order we keep only those edges $<e_1, e_2, ..., e_m>$ that have the sum of weights smaller than a certain threshold, computed as a percentage of the total sum of weights:

$$Sum_{G'} = \sum_{i=1}^{m} w_i < p\% * Sum_G = p\% * \sum_{i=1}^{n} w_i, \; p \in [0, 100], \; m <= n = |E|, \quad (1)$$

3.2 Deep Graph Convolutional Neural Networks

End-to-end deep learning architectures, such as the Deep Graph Convolutional Neural Network (DGCNN) [14], take as input graphs of arbitrary structure, (G, y), where y represents the label of the graph, and build a graph classification model by applying end-to-end gradient based training. As opposed to methods which use graph embeddings to transform graphs into tensor data that can be then classified via traditional machine learning algorithms, end-to-end methods solve a single joint optimization problem, which includes both feature learning and classification. This gives them the potential to produce better classification outcomes than the decoupled, embedding-based methods, but increases the complexity of the problem and thus, the computational effort needed to solve it. The DGCNN architecture is composed of three sequential parts. The first part extracts useful features for every node by using Graph Convolutional Networks (GCN). The extracted features characterize the graph topology and based on them, in the middle part, due to the use of the SortPooling layer, an ordering of graph nodes is defined. In the last part, the ordered sequence of nodes are introduced into a 1D convolutional neural network and then into dense layers with the purpose of learning a classification function. For a more in-depth description of the specific principles used by DGCNN, we refer the reader to [14].

3.3 DGCNN Interpretability

The next step after classifying graphs is to find the nodes which best discriminate between classes, in the attempt to interpret the model. We adapted Grad-CAM to graph classification models by starting from the premise that the graph nodes ordering resulting after the SortPooling layer in DGCNN encodes specific structural information (based on the relative structural relevance of the nodes within the graph), similar – in a way – to how pixels in neighboring regions of an image are correlated. In the following steps we detail our solution.

Let $F_1, F_2, ..., F_n$ be the feature maps in the final convolutional layer and S_c the score of the target class c. The corresponding gradients $(w_1, w_2, ..., w_n)$ are computed by using the formula:

$$w_i = \frac{\partial S_c}{\partial F}|F_i, \forall i = 1, .., n \tag{2}$$

These gradients are global-averaged pooled in order to obtain a weight of the importance of a feature map F_i for a target class c. By multiplying the weights w_i with their corresponding feature maps we obtain the weighted activations:

$$A_i = w_i * F_i, \forall i = 1, .., n. \tag{3}$$

The next step is to sum all the activations of the feature maps and apply the ReLU function, the result being a downsampled feature-importance array:

$$H = ReLU(\sum_{k=1}^{n} A_k) \tag{4}$$

We don't upsample H as it is done for images, we go back through the architecture to find an approximation of a group of nodes that are good predictors for a target class. The part of the architecture were we apply our reverse process is the CNN part, as depicted in Fig. 1. In this example, and even in the architecture which we used, this part is composed of two 1D convolutional layers and a Max-Pooling layer. The first 1D convolutional layer combines the resulted features of every node from the SortPooling layer into one feature. The dimensions of the ordered array does not change after this layer. Next, a MaxPooling operation is applied and, depending on the values of the hyperparameters, kernel and step, the dimensions of the previous array changes. A second convolutional layer is applied, changing the dimensions of the array again. We apply Grad-CAM on the result of the previous convolutional layer. Therefore, we can associate an element from H with a group of nodes, FG, by going back trough the architecture. In the illustrated example, $FG(1)$ (Final Group 1) is represented by two previous groups of nodes, where G1 (Group 1) contains the nodes 3 and 6, and G2 (Group2) contains the nodes 2 and 1. In the end, $FG(1)$ points to the nodes 3, 6, 2, 1. H consists of values between 0 (meaning that the group of nodes is not important in classifying the target class) and 1 (meaning that the group of nodes is a very good predictor for the target class). For every node, v_i, we discretize its importance into several bins, by defining an importance array, c_i, where the indices give us decimal intervals from H. For instance, index 0 represents the values between 0 and 0.1, index 1 represents values between 0.1 and 0.2, and so on. C is defined as a frequency matrix where the row c_i is the importance frequency array for the node v_i. This matrix is obtained by applying the *Importance frequency algorithm* (shown below) to every computed H. Based on the C matrix we then generate the interpretability heatmaps (Sect. 4.3) to visualize the discriminative nodes.

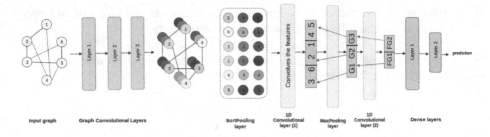

Fig. 1. DGCNN + Grad-CAM

Algorithm 1: Importance frequency algorithm

> **Input**   : H - importance array, FG - a list of lists of nodes
> **Output** : C - importance frequency matrix
> **Initialize:** $C(i) \leftarrow 0, \forall\, i = 1, ..., n, where\, n = |V|$
> 1 **foreach** *element $h \in H$* **do**
> 2   **foreach** *node $v \in FG(h)$* **do**
> 3 $idx \leftarrow \lfloor h * 10 \rfloor$;
> 4 $C(v)(idx) \leftarrow C(v)(idx) + 1$
> 5   **end**
> 6 **end**

---

4   Experimental Evaluation and Results

The domain specific problem we started off from consisted of graphs representing brain functional networks in two different physiological states. Though the classification accuracy obtained on that data was good, and the heatmaps obtained allowed for reaching a certain understanding of the generated models, we chose to validate the interpretability method more reliably, by generating artificial datasets in which the relation between the nodes in the graph is known in advance.

4.1   Data Generation

Validating interpretability methods for graph classification models is not straightforward, since if we employ real data it might not even be clear what the model should be learning. Because the interpretability model we propose tries to highlight class-relevant nodes, failing to do so may be caused by flaws in the interpretability model itself, but also by the fact that the classification model does not actually learn what it should. To remove the second factor from the analysis (since it is not relevant for the validation of the interpretability method), we turn to synthetic data to analyze the strengths and weaknesses of the proposed interpretation strategy. In generating the data, we followed three main objectives/hypotheses (further detailed in four data generation scenarios):

1. Classification performance on a random class distribution problem should be close to the 50% baseline – addressed by scenario $S1$ below; analyze what the interpretability heatmaps indicate in this situation.
2. Evaluate the robustness of the method to mild graph topologies and distributions which try to mimic the original, real-world problem we started from. This is addressed by generation scenarios $S2$ and $S3$ below.
3. Evaluate the robustness of the method to various complexities inherent in data, which normally affect performance, such as: (i) imbalance, (ii) overlap, (iii) noise and also combinations of these complexities (as most traditional machine learning techniques fail to handle well this aspect). Scenario $S4$ below considers two of these complexities.

The objectives above are materialized in the following 4 generation scenarios:

1. **Random classification problem (S1-Random)**, in which the graphs for both classes are very similar. We expect that the resulting model has very poor performance in this case (close to 50%), and the interpretability heatmaps to show no emerging pattern.
2. **Well separable and interpretable classification problem (S2-Easy)**, in which we select a subset of nodes to drive an almost perfect separation between classes. For this scenario, we expect classification accuracy to be close to 100% and the model to be able to learn which are the important nodes - which should be visible in the resulting interpretability heatmaps.
3. **Well separable, partially interpretable classification problem (S3-Moderate)**, in which we try to give more importance in separation to a subset of nodes, but this importance is not as straightforward as in the previous scenario. In this case, we also expect a very good classification accuracy, and the interpretability heatmaps should be able to indicate (at least partially) the important nodes.
4. **Partially separable, partially interpretable classification problem (S4-Hard)**, in which we attempt to make the data more difficult to separate, by introducing two challenges: imbalance and overlap.

The rest of this section describes the data generation processes for each of the above scenarios. All datasets generated contain 500 synthetic complete weighted, undirected graphs, each having 85 nodes. The graphs belong to 2 different classes - State1 and State2 - the class labels being uniformly distributed (except for S4-Hard, where we introduce imbalanced class distributions). Each node has the same labelling in all the graphs. The weights of the edges are numbers in the $[0, 1]$ range.

For **S1-Random**, the edge weights are drawn randomly from the same distribution for both classes, $\mathcal{N}_1(\mu_1, \sigma_1^2)$. As mentioned above, this should yield around 50% classification accuracy and the resulting heatmaps should not indicate any relevant nodes. For scenario **S2-Easy**, the graphs belonging to the first class are generated as for **S1-Random**; for the graphs belonging to the second class, we select a subset of k nodes for which we use a different edge weight distribution, $\mathcal{N}_2(\mu_2, \sigma_2^2)$. For the rest of the edge weights, we use the initial distribution, $\mathcal{N}_1(\mu_1, \sigma_1^2)$ - this should yield a (very) well separable classification problem. For this scenario we experimented with two different settings: one in which the weights of the k nodes in the separate community in *State2* were weaker than for the rest of the graph ($S2.1$), and one in which they were stronger ($S2.2$). The reason for this is to observe whether sparsification can affect class separability and model interpretability, since sparsification removes the smaller weight edges (thus it might remove relevant edges in $S2.1$). With **S3-Moderate** we tried to generate graphs that were separable by a well known network metric - the *betweeness centrality* - and see whether the model is able to learn those characteristics. More specifically, for the graphs belonging to the first class, we again generate complete, weighted graphs, drawing the weights randomly from $\mathcal{N}_1(\mu_1, \sigma_1^2)$. For the graphs belonging to the second class, we select a subset of k nodes and generate the weights of the edges connecting these nodes by drawing randomly from $\mathcal{N}_1(\mu_1, \sigma_1^2)$. Then, the rest of the nodes are "split" uniformly at random among these hub nodes. We thus create separated communities, within each community the edge weights being drawn randomly from $\mathcal{N}_2(\mu_2, \sigma_2^2)$. In a last generation step, we connect the nodes belonging to different communities (except for the hub nodes) by very weak connections, drawing their weights from $\mathcal{N}_3(\mu_3, \sigma_3^2)$. In **S4-Hard** we generate three different datasets. We keep the generation strategy from *S3-Moderate*, and try to make the classification problems harder by introducing first class imbalance, then class overlap. Dataset $S4.1$ was generated with an imbalance ratio of approximately 10, the second class being the minority class. For dataset $S4.2$, we employ a balanced class distribution but change the means of the three distributions used to generate the edge weights such as to make them overlap more. Finally, $S4.3$ was generated by applying jointly the strategies from $S4.1$ and $S4.2$.

The specific parameters for the distributions used are presented in Table 1. For k we experimented with three different values: 8, 42 and 77 for *S2-Easy*, and $k = 8$ for *S3-Moderate* and *S4-Hard*.

Table 1. Distributions used for data generation

*	$\mathcal{N}_1$		$\mathcal{N}_2$		$\mathcal{N}_3$	
	μ_1	σ_1	μ_2	σ_2	μ_3	σ_3
S1-Random	0.5	0.25	n.a.	n.a.	n.a.	n.a.
S2.1-Easy (weaker)	0.7	0.1	0.5	0.1	n.a.	n.a.
S2.2-Easy (stronger)	0.5	0.1	0.7	0.1	n.a.	n.a.
S3-Moderate	0.7	0.1	0.5	0.1	0.2	0.1
S4.1-Hard (imbalance)	0.7	0.1	0.5	0.1	0.2	0.1
S4.2-Hard (overlap)	0.6	0.1	0.5	0.1	0.4	0.1
S4.3-Hard (imb. + overlap)	0.6	0.1	0.5	0.1	0.4	0.1

4.2   Classification Performance Evaluation

The classification task was performed using the network structure for DGCNN as presented in [14], applied to input graphs sparsified to maintain a certain amount of edges, as specified by $p\%$. We repeated each experiment 10 times, using in each evaluation 80% of the data for training and the remaining 20% as validation (test) set. For setting $p\%$, we experimented with several options, from maintaining all edges (i.e. $p = 100\%$) down to keeping the strongest edges that make up for 50% of the total weights.

As expected, for *S1-Random*, the trained models learn to predict one of the classes, reaching an accuracy of around 50% (e.g. the average accuracy of the final model over the 10 runs for $p = 70\%$ sparsification threshold was 50.1%). For all the other scenarios, all models, in all runs, eventually converge to a 100% accuracy on the validation set. What differs is the speed of convergence and the variability of the accuracy on the validation set.

For example, if we compare the training behavior of the models in *S2-Easy* and *S3-Moderate* - see Fig. 2) - we observe that the latter converge faster, and with less variability, which might indicate that the models find these datasets easier to learn, contrary to our initial assumptions. A potential motivation for this can be found in the effect of sparsification. For *S3-Moderate* data, for the graphs belonging to *State2* we expect sparsification to remove the weak inter-community edges (i.e. the ones generated with $\mathcal{N}_3(\mu_3, \sigma_3^2)$). In contrast, for the graphs belonging to *State2* in *S2-Easy*, sparsification might remove edges from both outside and inside the community formed of the k nodes (with higher probability for the edges generated with the distribution having the lower mean); what is important to note here is that by removing from both outside and inside the community, the problem might become more difficult to learn. Within the same scenario, we observed that keeping more edges in the initial graphs (by increasing $p\%$) makes the models converge more slowly (i.e. in later epochs), which was expected.

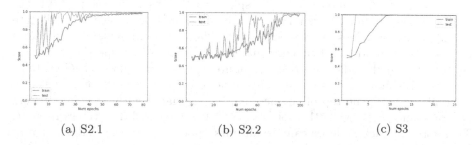

(a) S2.1   (b) S2.2   (c) S3

Fig. 2. Comparison of learning curves for *S2-Easy* and *S3-Moderate*, $k = 8$, $p = 70\%$ (Score = Average Accuracy)

In *S4-Hard* we find that the models have different learning patterns according to the complexities added to the data (imbalance and/or overlap). As illustrated in Fig. 3a, the model only starts to learn something meaningful around epoch 25, when training accuracy starts to increase from 90% towards 100% (before that epoch, the model always predicted class *State1*, also reflected in the value of the validation set accuracy). We observe a similar behavior in Fig. 3c, only this increase appears later in training, due to the overlap also being a data complexity that the model has to overcome. Comparing with the behavior in *S3-Moderate* (see Fig. 2c), we find that both imbalance and overlap make the models learn more slowly, and overlap induces more variability in the learning process (which can be seen in the validation/test set accuracy).

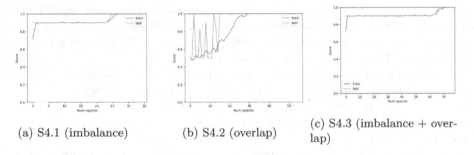

(a) S4.1 (imbalance)   (b) S4.2 (overlap)   (c) S4.3 (imbalance + overlap)

Fig. 3. Comparison of learning curves for *S4-Hard*, $k = 8$, $p = 70\%$ (Score = Average Accuracy)

4.3   Interpretability Heatmaps

In order to visualize the discriminative nodes we created a heatmap where the horizontal axis (Oy) represents the nodes of the graph, G, and the vertical axis (Ox) represents the interval $[0, 1]$, the importance of a node in classification (the values from H). The interval $[0, 1]$ is split into 10 bins with a step of 0.1: the values between 0 and 0.1 are removed for better visualization, while we allocate an extra bin for values exactly equal to 1. The color represents the difference between the importance frequency matrix of *State1*, C_1, respectively *State2*, C_2 ($C1 - C2$), the difference being computed per decimal interval. C contains values which indicate how many times a node takes a value from H within a decimal interval. The top of the heatmap is associated with high values from H (for example, 1), while the bottom of the heatmap with low values from H. Therefore, for each node we can visualize its importance in the classification as follows: if the red color appears on the top of the image, and the blue color on the bottom of the image, it means that the node is a good predictor for *State1*; if we have the blue color on the top of the image and the red color on the bottom, it means that the node is very important in classifying *State2*; the green colour shows that the node does not have discriminative power in the model; if red or blue colors appear emphasized only in the middle of the heatmap, it might indicate that our problem is difficult to learn, the difference between classes being less noticeable. We created an average heatmap across the 10 folds in order to capture the strongest common features of the models resulting from different evaluation folds.

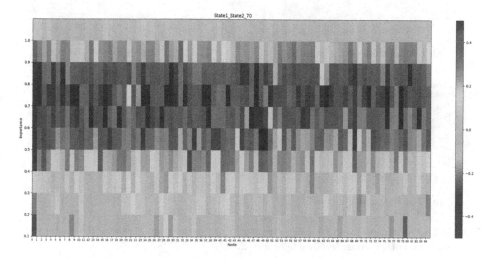

Fig. 4. The heatmap for *S1-Random* data generation strategy where $p = 70\%$.

Figure 4 illustrates the average heatmap for the models learned for *S1-Random*. The model predicts any graph in the test/validation set as belonging to the *State1* class; the heatmap indicates that all nodes in the graph are relevant for predicting that class, which is to be expected.

For *S2.1-Easy (weaker)* data generation strategy we performed experiments and computed heatmaps for the following sparsification percentages: $p = 100\%$, $p = 70\%$ and $p = 50\%$. The purpose of this experiment was to highlight the k nodes whose edge weights were generated using a different distribution in *State2*. We always choose the k nodes to be the first in the graph (i.e. the leftmost 8 columns of the heatmap in Fig. 5). In Fig. 5a, we notice that if we keep all the edges no clear patterns emerge, because the information is actually distributed across the nodes. But if we sparsify the graphs using a percentage $p = 70\%$, Fig. 5b shows how our classifier distinguishes between classes by highlighting the 8 nodes that are good predictors for *State2* (and the model performance is almost the same). If we sparsify more, $p = 50\%$, Fig. 5c does not indicate clear patterns because through sparsification the nodes are losing their importance (for example, the discriminative edges are eliminated).

In a next experiment, we modified k, the number of nodes for which we employed a different distribution for generating the edge weights (for the graphs belonging to the second class). The results for $p = 70\%$ can be visualized in Fig. 6. Figure 6a illustrates a clear difference between the discriminative nodes

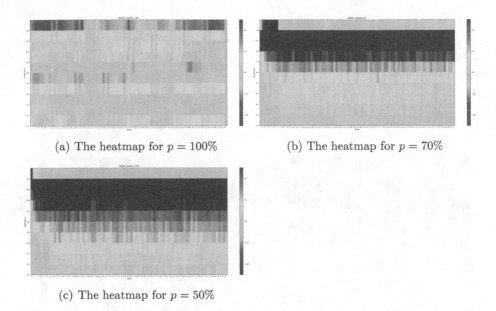

(a) The heatmap for $p = 100\%$ (b) The heatmap for $p = 70\%$

(c) The heatmap for $p = 50\%$

Fig. 5. The heatmaps for *S2.1-Easy (weaker)* data generation strategy for each sparsification percentage considered, where $k = 8$.

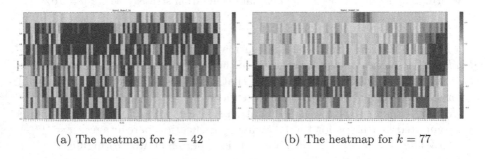

(a) The heatmap for $k = 42$ (b) The heatmap for $k = 77$

Fig. 6. The heatmaps for *S2.1-Easy (weaker)* data generation strategy, at $p = 70\%$.

for *State*1 (left part), and the good predictors for *State*2 (right part). Theoretically, by increasing k, we should have more discriminative nodes for one state. Figure 6b shows the opposite: actually the nodes (left part) whose edges weights have not been generated from another distribution are the most important ones for *State*1. Also, we can notice that there are fewer good predictors (the nodes from the middle of the heatmap) for *State*2 than in the previous case when $k = 42$. In the case of *S2.2-Easy (stronger)*, the first 8 modified nodes are more important in classifying *State*1 rather than *State*2 as it is shown in Fig. 7a, while in Fig. 7b all nodes are relevant in classifying *State*1.

In *S3-Moderate* the first 8 nodes were selected to be the hubs in *State2*. As Fig. 8 shows, only a part of them are highlighted as being important in classifying *State2*.

Even though *S4.1-Hard* represents a class imbalance problem, the same patterns as in *S3-Moderate* emerge in Fig. 9a, only the colors are less intense, which might indicate that the model is less certain in how the two classes separate. A similar phenomenon can be observed when the two classes overlap more, in *S4.2-Hard* (Fig. 9b), where the strong shades of blue and red appear more towards the middle bins (as opposed to the top or bottom of the heatmap - as for models which converged faster and are - intuitively - more confident in their separation). An interesting phenomenon can be observed for *S4.3-Hard* (Fig. 9c), where, as expected, the emphasized patterns appear in the middle of the heatmap, but the heatmap is flipped (blue appears more on the top of the heatmap, while red more on the bottom part).

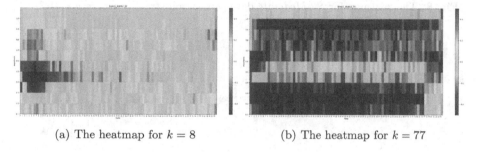

(a) The heatmap for $k = 8$ (b) The heatmap for $k = 77$

Fig. 7. The heatmaps for *S2.2-Easy (stronger)* data generation strategy for $k = 8$, respectively $k = 77$, where $p = 70\%$

5   Discussion

The interpretability method proposed in this paper attempts to extract information about the importance of graph nodes in achieving class separation for deep graph convolutional models. The evaluation attempted to assess the validity of the method on several classification tasks for which - intuitively - we know what to expect from the models. A first important observation is that sparsification affects the outcome of the interpretability method, and this is because it affects how the underlying classification model learns to separate between the classes. When the information is dense (i.e. we keep all graph edges), individual nodes matter less in learning how to separate between the classes - which is to be expected. Naturally, the "right" amount of sparsification is highly dependent on the problem, and - even if not observed in the current evaluations - sparsification affects not only interpretability, but also the classification performance. Consequently, a future step is to study these interactions more systematically.

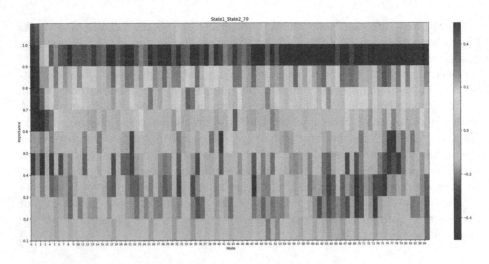

Fig. 8. The heatmap for *S3-Moderate* data generation strategy where $p = 70\%$.

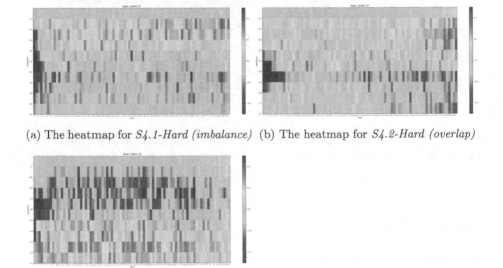

(a) The heatmap for *S4.1-Hard (imbalance)*   (b) The heatmap for *S4.2-Hard (overlap)*

(c) The heatmap for *S4.3-Hard (imb + overlap)*

Fig. 9. The heatmaps for *S.4-Hard* data generation strategy where $p = 70\%$.

By comparing the heatmaps for *S3-Moderate* and *S4-Hard*, and considering also how the corresponding models converge, we believe that the heatmaps may capture also the confidence of the model's predictions. However, this phenomenon needs to be studied further, especially for classification problems which are not perfectly separable.

The proposed modification to Grad-CAM performs a very rough approximation to compute graph node relevance. We are currently exploring more accurate alternatives for doing this (such as adapting the deconvolution method initially proposed for the interpretation of image convolutional models).

6   Conclusion

Interpretability is – in many application domains – crucial towards gaining acceptance for machine learning models. Graph convolutional models add an extra layer of difficulty for interpretability methods, because graphs do not exhibit clear spatial relations between their nodes (like images do).

In this paper we propose a method for graph classification and model interpretation, which combines DGCNN with a modified Grad-CAM algorithm, to obtain heatmaps representing each node's relevance to the classification of a specific graph. We alter the Grad-CAM algorithm to apply only operations which do not assume a specific locality for nodes. We evaluate our method on synthetic datasets which were generated to emulate a real dataset representing brain functional networks in different physiological states. These functional networks are represented by complete, weighted graphs that need to be sparsified. The resulting heatmaps are generally able to identify the nodes which we intended to be relevant for the identification of a specific class. Interrestingly, we believe they manage to also capture some degree of "uncertainty" associated to the predictions of the model, but this aspect needs further investigation, together with the effect of sparsification on the resulting models and heatmaps.

Acknowledgments. This work was supported by a grant from the Romanian National Authority for Scientific Research and Innovation, CNCS-UEFISCDI (project number COFUND-NEURON-NMDAR-PSY), a grant by the European Union's Horizon 2020 research and innovation program – grant agreement no. 668863-SyBil-AA, and a National Science Foundation grant NSF-IOS-1656830 funded by the US Government.

References

1. Doshi-Velez, F., Kim, B.: Towards a rigorous science of interpretable machine learning. arXiv e-prints, February 2017
2. Selvaraju, R.R., Das, A., Vedantam, R., Cogswell, M., Parikh, D., Batra, D.: Grad-CAM: why did you say that? Visual explanations from deep networks via gradient-based localization. CoRR abs/1610.02391 (2016)
3. Molnar, C.: Interpretable machine learning (2019). https://christophm.github.io/interpretable-ml-book/
4. Greenwell, B.M., Boehmke, B.C., McCarthy, A.J.: A simple and effective model-based variable importance measure (2018)
5. Zhao, Q., Hastie, T.: Causal interpretations of black-box models (2019)
6. Fisher, A., Rudin, C., Dominici, F.: All models are wrong but many are useful: variable importance for black-box, proprietary, or misspecified prediction models, using model class reliance. arXiv e-prints, January 2018. arXiv:1801.01489

7. Goldstein, A., Kapelner, A., Bleich, J., Pitkin, E.: Peeking inside the black box: visualizing statistical learning with plots of individual conditional expectation. J. Comput. Graph. Stat. **24**(1), 44–65 (2015)
8. Štrumbelj, E., Kononenko, I.: Explaining prediction models and individual predictions with feature contributions. Knowl. Inf. Syst. **41**(3), 647–665 (2014)
9. Ribeiro, M.T., Singh, S., Guestrin, C.: "Why should I trust you?": explaining the predictions of any classifier. CoRR abs/1602.04938 (2016)
10. Zeiler, M.D., Fergus, R.: Visualizing and understanding convolutional networks. CoRR abs/1311.2901 (2013)
11. Simonyan, K., Vedaldi, A., Zisserman, A.: Deep inside convolutional networks: visualising image classification models and saliency maps. CoRR abs/1312.6034 (2013)
12. Springenberg, J., Dosovitskiy, A., Brox, T., Riedmiller, M.: Striving for simplicity: the all convolutional net. In: ICLR (Workshop Track) (2015)
13. Zhou, B., Khosla, A., Lapedriza, A., Oliva, A., Torralba, A.: Learning deep features for discriminative localization. In: CVPR (2016)
14. Zhang, M., Cui, Z., Neumann, M., Chen, Y.: An end-to-end deep learning architecture for graph classification. In: AAAI, pp. 4438–4445 (2018)

Streams and Times Series

Soft Voting Windowing Ensembles for Learning from Partially Labelled Streams

Sean L. A. Floyd and Herna L. Viktor[✉]

School of Electrical Engineering and Computer Science, University of Ottawa,
Ottawa, ON, Canada
{sfloy029,hviktor}@uottawa.ca

Abstract. Mining data streams has become an important topic due to the increased availability of vast amounts of online data. In such incremental learning scenarios, observations arrive in a sequence over time and are subject to changes in data distributions, also known as concept drifts. Interleaved test-then-train evaluations are often used during supervised learning from streaming data. The idea is intuitive: we first use each instance to test a model, then it is used for training. However, true class labels may be missing or arrive well after the prediction, which implies that they cannot be used for training and/or drift detection. Based on these considerations, we introduce our LESS-TWE ensemble-based method for online learning in domains where full reliance on labels would be unfeasible. Our approach combines weighted soft voting and unsupervised drift detection to reduce the dependency on labels during model construction. In cases where the label is unavailable, the most confident label, as predicted through weighted soft voting, is selected. Similarly, our unlabelled drift detector flags for drifts based on the voting confidence, rather than relying on the true label. Our experimental evaluation indicates that our algorithm is very fast, achieves comparable predictive accuracy when compared to the state-of-the-art and outperforms baseline methods.

Keywords: Stream mining · Prequential evaluation · Unlabelled data · Ensembles · Self-training · Unsupervised drift detection

1 Introduction

A major challenge in learning from data streams is late-arriving or missing class labels. Assuming that data will arrive correctly labelled and in a timely manner often does not reflect reality, and, as such, limits the applicability of supervised methods [11]. Many real-world problems, in areas such as cybersecurity and fraud detection, require the use of semi-supervised techniques to handle partial, and potentially sporadic, lack of labels. Furthermore, the unavailability of labels means that the interleaved test-then-train paradigm, which is common practice in online learning evaluation [17], cannot be followed. Online learning algorithms

© Springer Nature Switzerland AG 2020
M. Ceci et al. (Eds.): NFMCP 2019, LNAI 11948, pp. 85–99, 2020.
https://doi.org/10.1007/978-3-030-48861-1_6

need to adapt their models to potential changes in the underlying concepts. To this end, techniques have been developed to explicitly detect these changes allowing algorithms to adapt models more quickly. Drift detection methods rely mainly on the use of labelled data by detecting changes in the accuracy of a classifier over time [1, 7] or use some form of statistical tests [14].

Ensembles have shown to obtain superior performance by increasing the accuracy and diversity of single classifiers, both in the offline and online settings [3, 10]. Online ensemble learning from partially labelled streams has, to date, received very limited attention. Current techniques generally rely on first clustering unlabelled data, which is computationally expensive and may have limited applicability in domains where just-in-time models are required [8, 11]. The goal of this paper is to narrow the gap in research as it pertains to ensemble learning of evolving streams where the labels are limited, or their on-time arrival cannot be guaranteed. Our aim is to follow an interleaved test-then-train paradigm, as is standard practice in the domain, while reducing the dependence on true labels. In addition, our work eliminates the need for initial unsupervised learning.

Our Learning from Evolving Streams via Self-Training Windowing Ensembles (LESS-TWE) framework employs a novel windowing technique with a weighted soft voting strategy. In our self-training stage, the ensemble's confidence is used to predict the label when unavailable. In addition, we introduce an unsupervised drift detection algorithm that extends the Fast Hoeffding Drift Detection Method for evolving Streams (FHDDMS) algorithm [13], to further reduce the reliance on class labels. By utilising a hybrid sliding-tumbling windows technique for instance selection, where the instances seen by individual classifiers are interleaved, we aim to achieve savings relating to the execution time, while maintaining training diversity and high predictive accuracy.

This paper is organised as follows. Section 2 introduces background concepts. We present our LESS-TWE framework in Sect. 3. In Sect. 4, we describe our experimental evaluation. Finally, in Sect. 5 we conclude the paper and discuss our future research.

2 Background

Online learning from data streams differ from batch learning in several important ways. Firstly, since we have continuous flow of data, models need to be built and updated as the instances arrive, using limited memory and time. Secondly, due to changes in data distributions, also known as concept drifts, the models must be swiftly able to detect and to adapt, to maintain high accuracy [17].

Traditional metrics based on error rate can be misleading if used as a proxy for evaluating algorithms for evolving streams. Interleaved test-then-train methods ensure that the model has not previously seen new test instances, implying that no holdout test set is necessary [17]. This procedure assures that a classification model is being tested on unseen instances as the stream evolves. In contrast to the holdout procedure, it makes the maximum and immediate use of data, meaning all instances are used for both testing and training. It also ensures a smooth plot of accuracy over time, as the impact of each example becomes increasingly less significant to the overall average. In this procedure, the accuracy of a learning algorithm is punished for early mistakes. This effect, however, diminishes after training thousands of instances over time. As more data is tested in the test-then-train approach than in the holdout method, each instance used to assess the accuracy and performance of the model weighs less than it would have in a smaller holdout test set [3]. As such, interleaved test-then-train evaluation has become standard practice in online learning research, including when assessing online ensembles.

Online ensembles for supervised learning from evolving stream, i.e. streams that are non-stationary and are susceptible to concept drift, has resulted in success in many domains. In an ensemble, the notion of diversity ensures that the combined vote exceeds that of the individual committee members. Typically, each component classifier either focuses differently on (often weighted) subsets of the instances, or uses diverse feature subsets when constructing its model. [10] provides a detailed review of online ensembles for evolving streams, including both chunk-based and instance-based methods, such as Adaptive Classifier Ensembles (ACE) [12], Leveraged Bagging (LB) [2] and Streaming Ensemble Algorithm (SEA) [15]. While successful, these supervised algorithms rely on instantly available labels, to maintain high accuracy and to flag for drift, making them unsuitable for direct use in a setting where the full reliance on labels cannot be ensured. In the next section, we introduce our LESS-TWE approach for learning from such partially labelled streams.

3 LESS-TWE Online Learning

This section presents the details of our LESS-TWE online learning algorithm. Figure 1 illustrates how our contributions fit together and operate in one iteration of an interleaved test-then-train loop. Note that instances may arrive in chunks or one at a time.

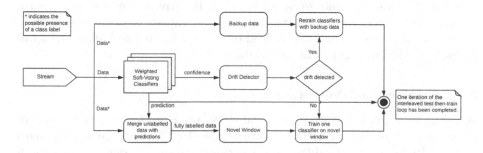

Fig. 1. High-level overview of the LESS-TWE methodology

Algorithm 1: LESS-TWE algorithm

---

1 **while** *stream.has_more_instances()* **do**
2 $\quad$ X, y = stream.get_next_tuples(number_of_instances_to_fetch);
3 $\quad$ predictions, probabilities = voting_ensemble.predict(X);
4 $\quad$ drift_detected = voting_ensemble.detect_drift(predictions, probabilities);
5 $\quad$ **if** *true_label_percentage* $\neq$ 100 **then**
6 $\quad\quad$ **if** *drift not detected* **then**
7 $\quad\quad\quad$ y = label_with_predictions(y, predictions, true_label_percentage);
8 $\quad$ voting_ensemble.train(X, y);
$\quad$ // This algorithm shows the steps required to modify learning when
$\quad$ there are unlabelled data
9 **function** *label_with_predictions(y, predictions, true_label_percentage)*
10 $\quad$ **for** *index* = 0 ; *index* < *length(y)* ; *index*+ = 1 **do**
11 $\quad\quad$ **if** *random_number_between(0, 100)* $\lneq$ *true_label_percentage* **then**
12 $\quad\quad\quad$ y[index] = predictions[index];
13 $\quad$ **return** *y*;

3.1 Hybrid Sliding-Tumbling Windows

Our ensemble is designed so that the training sets of the N individual classifiers proceed out of step, using tumbling windows for instance selection. At every iteration of the interleaved test-then-train loop, we append the new tuples to the ensemble's window and train a *single* classifier in the ensemble on that window. For the next $N - 1$ iterations, we train the remaining $N - 1$ classifiers, and so on. We do this so that from the point of view of the ensemble, we are using sliding windows to train. However, from the point of view of each classifier in the ensemble, we are employing tumbling windows to train, where the classifiers proceed out of step. Figure 2 illustrates this method for three (3) classifiers in an ensemble where the difference in starting points of classifiers is equal to 1 instance. Each classifier will learn only from the same coloured batch; (as seen in Fig. 2) classifier c_1 will train on the purple window, while classifiers c_2 and c_3 will utilise the blue and green windows [5]. Intuitively, this method ensures diversity, in terms of the instances used by the individual classifiers when casting their votes, and has the potential to lead to savings, in time and memory, as will be indicated in our experimental evaluation.

3.2 Weighted Soft Voting

When an ensemble determines a class label for a new instance, each component classifier first returns its individual prediction [10]. The ensemble then maps these multiple, potentially different, predictions to a single value. To accomplish this, several functions have been proposed that apply weights to any permutation of the classifiers and/or of their predictions, and then apply a combination rule to these values to finally vote on the final output of the ensemble.

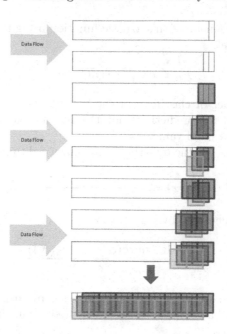

Fig. 2. Sliding tumbling windows [15]. (Color figure online)

Soft voting requires each of the classifiers in the ensemble to output a confidence value (usually in $[0,1]$) for their prediction for each class value or to output the probabilities that an instance belongs to a given class label. In the case of *simple soft voting*, the average probability for each class label is computed over the predictions of all classifiers. The probability of the final class label is given by Eq. 1. Here, $h_i^j(x) \in [0,1]$ takes value 1 if classifier h_i predicts class label c_j. L is the set of class labels, where l is any label in L.

$$H(x) = max(\frac{1}{n} \sum_{i=1}^{n} h_i^l(x) \ \forall l \in L) \tag{1}$$

Our LESS-TWE ensemble extends the simple soft voting method by employing two weighted soft voting schemes, each of which outputs a probability or confidence in the predicted label. We employ weighting schemes to determine whether distinguishing between small variations in votes may potentially create more diversity in our online learning setting and correct misclassifications [2]. In our voting schemes, the label with the highest average value is selected and the labelled instance is added to the training data.

In our first variant, we use the logistic sigmoid activation function defined by:

$$f(x) = \frac{L}{1 + e^{-k(x-x_0)}} \tag{2}$$

where e is the natural logarithm base, x_0 is the x-value of the sigmoid's midpoint, L is the curve's maximum value, and k is the steepness of the curve. This voting strategy computes the mean of this logistic function using the prediction of each classifier for each label:

$$\frac{1}{n} \sum_{i=1}^{n} \frac{1}{1 + e^{-k(p_i(X)-\gamma)}} \tag{3}$$

where n is the number of classifiers in the voting ensemble, and $p_i(X)$ is the probability of classifier i predicting that the instance in question belongs to class X and γ is a threshold value.

The second variant use the hyperbolic tangent weighting function defined as:

$$f(x) = \frac{1 - tanh(\alpha - \beta x)}{\delta} \tag{4}$$

and the sum is calculated by

$$\frac{1}{n} \sum_{i=1}^{n} \frac{1 - tanh(\alpha - \beta \times p_i(X))}{\delta} \tag{5}$$

It follows that the values of α, β and δ are set by inspection. The average is used to compute a base learner's true label prediction, similar to the logistic sigmoid voting scheme.

3.3   Online Labelling

Our online labelling method is based on the self-training semi-supervised learning paradigm. In offline self-training, an initial hypothesis is learned from labelled data, which is then used to classify unlabelled data. The most confident unlabelled data, along with the labelled data, are added into the training set, which is used repeatedly to refine the hypothesis. Thus, selective self-training employs the most confident of a classifier's predictions to label the unlabelled data [16] and is a promising way to deal with unlabelled data in an offline semi-supervised learning setting. However, self-training can reinforce classification errors and may lead to reduced accuracy [16]. Also, in an online setting where we employ the interleaved test-then-train paradigm, the availability of all true labels cannot be guaranteed. We therefore need to modify the algorithm for online learning. To this end, our online labelling algorithm 1 uses predictions for data arriving without. This algorithm combines the most confident predictions from the individual classifiers to increase the overall confidence in the proposed label. Intuitively, the windows of instances seen by the N different classifiers in our ensemble differ at any point in time, and such diversity should lead to more robust results.

Specifically, as an instance X arrives, it is first tested and next used for training. If X's true class label y is known upon arrival, then y is used for training. Alternatively, if X arrives without a label, the ensemble votes to assign the predicted class label $\hat{y}$ with the highest confidence, which is subsequently used for training. This process continues indefinitely, until the end of the stream, unless concept drift is detected. In cases where concept drift is detected, the ensemble is reset, and a new interleaved test-then-train cycle begins.

3.4 Unlabelled Drift Detection

Our drift detection approach relies on storing, in sliding windows, whether a classifier correctly predicts a label (0 or 1). If the number of incorrect predictions is higher than a threshold t, then the algorithm flags for a drift. In our work, we modified the FHDDM method [13] so that the dependence on labels is removed and we now determine the label of unlabelled instances using a probability. Thus, when predicting the class, each classifier in the ensemble outputs a probability for a label being correct. For example, a classifier could output the following class probabilities for a three-class task $\{A : 0.24, B : 0.11, C : 0.65\}$ and these values are used in our calculation.

Algorithm 2: Modified Fast Hoeffding Drift Detection Method (MFHDDM)

---

1 **function** *init(window_size, delta, use_probability)*
2   (n, δ, p) = (window_size, delta, use_probability);
3      $\epsilon_d = \sqrt{\frac{1}{2n} ln \frac{1}{\delta}}$;
4   reset();
5 **function** *reset()*
6   w=[];
7      $\mu^m = 0$;
8 **function** *detect(p)*
9   **if** $w.size = n$ **then**
10   w.tail.drop();
11   w.push(p);
12   **if** $w.size < n$ **then**
13   return False;
14   **else**
15         $\mu^t = w.average()$;
16   **if** $\mu^m < \mu^t$ **then**
17            $\mu^m = \mu^t$;
18         $\Delta\mu = \mu^m - \mu^t$;
19   **if** $\Delta\mu \geq \epsilon_d$ **then**
20   reset();
21   return True;
22   **else**
23   return False;

Algorithm 2 shows our drift detection approach which calculates the *average* probability for the winning class as obtained from the members of the ensemble. We maintain two sliding windows, a short window to detect abrupt drift, and a long window for detecting gradual drift. Following [13], Hoeffding's bound is used to detect if the average probability drifts too far from the maximum seen average probability using the following test.

FHDDM Test [13]: In a stream setting, assume μ^t is the mean of a sequence of n random entries, where the prediction status of each instance is represented by a value in the set $\{0, 1\}$, at time t. Let μ^m be the maximum mean observed so far. Let $\Delta\mu = \mu^m - \mu^t \geq 0$ be the difference between the two means. Given the desired δ, Hoeffding's inequality implies that a drift has occurred if $\Delta\mu \geq \varepsilon_d$, where:

$$\varepsilon_d = \sqrt{\frac{1}{2n} \ln \frac{1}{\delta}} \tag{6}$$

If a drift is detected, the ensemble is reset and the interleaved test-then-train loop recommences.

4   Experimental Evaluation

All experiments were conducted on a MacBook Pro base model 11,4, running macOS 10.14.4. The framework's implementation and the code for the experiments are available at GitHub[1]. This repository also shows the results of extensive experiments in terms of parameter combinations including window types and sizes, batch sizes, base learners (Hoeffding trees (HT), Multinomial and Gaussian Naive Bayes (NB) and Stochastic Gradient Descent (SGD)) and drift detection methods. Our initial experimental evaluation confirms that the hyberbolic tangent weighting function is computationally much more efficient (in general, at least 4 times faster) than the logistic sigmoid scheme is; we thus use this voting method throughout our evaluations.

The no-change classifier as well as a majority-class classifier were used as our baselines. The no-change classifier assumes that the class label of instance k would be the same as instance $k - 1$, while the majority class learner assigns the class seen most often, so far, to instance k. We further compare our LESS-TWE ensemble to the previously mentioned, state-of-the-art leveraged bagging (LB) algorithm, using parameter combinations selected by inspection after having been ranked with averaging Eq. 7. The LB classifier is implemented with a built-in ADWIN drift detector that replaces the worst performing classifier in the ensemble with a new classifier when a drift is detected [3]. Bifet et. al. [2] postulated that adding more random weights to all instances seems to improve accuracy more than if only adding to the misclassified instances, as is common in traditional bagging methods. For this reason, [2] proposed the online LB algorithm with randomisation improvements, namely increasing the weights of the input samples and adding randomisation to the output of the ensemble via output codes. We have chosen the LB technique in our comparison, since it is a robust framework for classifying evolving streams that has shown to yield highly accurate results through increasing the input-space diversity. Also, the randomisation approach followed by the LB method is related to the weighted soft voting scheme we use in the LESS-TWE approach.

---

[1] https://github.com/SeanLF/scikit-multiflow.

Recall that we use the interleaved test-then-train evaluation method in all our experiments. The performance measures are the execution time (measured in seconds), as well as the κ_t statistic to evaluate a classifier's predictive performance. This κ statistic compares our classifier to a no-change classifier and takes temporal dependence in the data into account [17].

4.1 Benchmark Data Sets

The benchmark data sets used for our analysis are SEA [15], CIRCLES, SINE1 and MIXED, which all contain noise, and either abrupt or gradual concept drifts. We have generated data for SEA, while the data generated for the last three data sets were obtained from [13]. Each experiment is run on five different examples each sourced from three synthetic data sets (5 examples of SINE1, another 5 of CIRCLES, etc.) and three streams generated by the SEA generator with levels of noise in increments of ten percent (from 0% to 20%), for a total of eighteen (18) streams of one hundred thousand (100,000) instances.

The **CIRCLES** data stream is composed of two relevant numerical attributes: x and y, which are uniformly distributed in $[0, 1]$. There are four concepts in this data set, each representing whether a point is within a circle given x and y coordinates for its centre and its radius r_c. This data set contains gradual concept drifts that occur at every twenty-five thousand (25,000) instances.

SINE1 contains abrupt concept drifts. It has only two relevant numerical attributes, for which the values are uniformly distributed in $[0, 1]$. Before the concept drift, all instances for values below the curve $y = sin(x)$ are classified as **positive**. Then, after the concept drift, the rule is reversed; therefore, the values below the curve become **negative**. The drifts were generated at every twenty thousand (20,000) instances.

MIXED also contains abrupt concept drifts and uses four relevant attributes, two of which are Boolean, let them be v and w; and the other two attributes are numerical, in $[0, 1]$. Instances belong to the positive class if two of three conditions are being met: v is true, w is true, $y < 0.5 + 0.3 \times sin(3\pi x)$. For each concept drift, the conditions are reversed, meaning that if the conditions are met, it will be a positive instance, then after the drift, it will be a negative instance. The abrupt concept drifts occur at every twenty thousand (20,000) instances.

SEA generates streams with abrupt concept drift, composed of three numerical attributes of values in $[0, 10]$, where only the first two attributes are relevant. For each instance, the class is determined by checking if the sum of the two relevant attributes passes a threshold value. Let f_1 and f_2 be the two numerical relevant attributes, and θ the threshold. An instance belongs to class *one* if $f_1 + f_2 \leq \theta$. As in [15], our stream has four concepts, with the threshold values for each being 8, 9, 7 and 9.5. We generate streams of one hundred thousand (100,000) instances, from zero to twenty percent noise, in ten percent increments ($\{0; 10; 20\%\}$). Drifts, therefore, occur at every twenty-five (25,000) thousand instances.

4.2   Effects of Training with a Lower Percentage of Labelled Data

In this section, we aim to determine the percentage of labelled data used at which our LESS-TWE ensemble's κ_t metric declines. The parameter combinations were selected by inspection after having been ranked with an averaging equation (7) that assigns more weight to the rank for the κ_t metric than the execution time. The ranking algorithm from the Nemenyi test was extracted to rank the possible parameter combinations, with $\delta = 3$ and $\chi = 4$.

$$\frac{\delta \times rank_{\kappa_t} + rank_{execution\ time}}{\chi} \tag{7}$$

Table 1 lists the global predictive accuracy and the κ_t values for each of the benchmarking data set. The table indicates that the global predictive accuracy of our ensemble is not significantly reduced by training with less ground truth. However, the predictive performance of the ensemble, as measured by κ_t, differs between 20% and 54% when examining κ_t when using 60% and 100% of ground truth.

Table 1. Accuracy (%) and κ_t when training with varying percentages of labelled data.

GT	CIRCLES		MIXED		SINE1		SEA 0%		SEA 10%		SEA 20%	
	Acc	κ_t	Acc	κ_t	Acc	κ_t	Acc	κ_t	Acc	κ_t	Acc	κ_t
100	78	0.56	80	0.60	84	0.69	97	0.94	82	0.62	70	0.38
90	76	0.52	79	0.59	84	0.68	97	0.94	82	0.62	70	0.38
80	73	0.46	79	0.58	82	0.65	96	0.93	82	0.62	70	0.39
70	68	0.36	77	0.55	80	0.60	96	0.92	82	0.62	71	0.40
60	63	0.26	72	0.44	77	0.55	95	0.90	83	0.64	70	0.39

From Table 1, we find that our ensemble does not suffer a drastic reduction in its global predictive accuracy when training with only 60% ground truth (in other words, training with 40% less labelled data). However, the κ_t statistic indicates mixed results. For the experiments using the SEA data sets, using less labels do significantly reduce the accuracy. Additionally, one notes that the CIRCLES data set is hard to model with access to a lower percentage of labelled data, which is perfectly logical given what the class label represents. Indeed, very specific data instances are required to model the class well, given that it represents whether a data instance resides within the radius of a predefined circle. Results show clearly that the reduction in κ_t is highly dependent on the data set being modelled, as should be expected.

4.3   Comparison in Terms of Accuracy and Runtime

Figure 3 shows the results when we compare our LESS-TWE framework with the LB technique, a single HT learner and the SGD classifier, as well as the no change and majority class classifiers. In this set of experiments, we include various configurations of LESS-TWE ensembles where we vary the percentages of ground truths, the window types and sizes, the base learner as well as the weighted soft voting scheme. The results show that the LB algorithm consistently performed well, in terms of κ_t values, with LESS-TWE ensembles consistently in second place. Our results also indicate that the no change and majority class learners produced low κ_t values, while the HT method appears to be sensitive to noise. That is, these algorithms were unable to learn the concepts well and to adapt to drifts or noise.

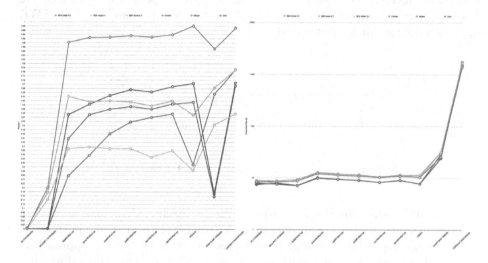

Fig. 3. Comparative κ_t and execution times across multiple algorithms.

To determine which pairs of algorithms actually differ, we converted the results into a ranking task [6,9] and used the post-hoc Nemenyi test as shown in Fig. 4, where a lower rank means a better predictive accuracy (a better κ_t). The graph shows that there is no significant statistical difference among the LB method, our LESS-TWE ensemble using our best overall parameter combination, an SGD classifier, and our LESS-TWE ensemble using our hybrid windowing approach. This supports the rejection of the null hypothesis for κ_t. Note that there is a significant statistical difference between both the majority voting and no change classifiers with all algorithms, aside from our LESS-TWE ensemble training with 70% of ground truth or less. Therefore, this test showed that our LESS-TWE ensemble, using our preferred parameter combinations, performed comparably to the LB approach, when the number of labels is adequate for the problem domain.

Considering execution time, we again used the post-hoc Nemenyi test to determine which pairs of algorithms differ as shown in Fig. 5. The graph indicates that the LB ensemble ranks last, and the HT decision tree ranks second to

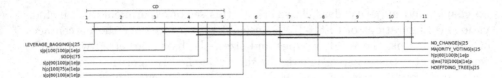

Fig. 4. Nemenyi graph ranking κ_t for various algorithms

last, as shown in Fig. 3. The graph also illustrates that there is a significant statistical difference between the LB and LESS-TWE ensembles. Given that the LB algorithm runs in over two orders of magnitude longer than the LESS-TWE approach, this result is as expected.

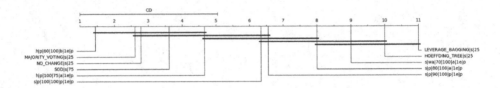

Fig. 5. Nemenyi graph ranking execution times for various algorithms

LESS-TWE ensembles clearly outperforms the baseline algorithms against all streams. Our results further show that the LESS-TWE approach is much faster than the LB algorithm and that it performs on par with the LB approach in terms of the κ_t metric. Our results also indicate that the predictive performance of our ensemble when trained with 80% and 90% ground truths do not present significant differences to that of the LB algorithm. Our LESS-TWE algorithm brings outstanding time savings in algorithm execution-time, running approximately 160 times faster than the LB method does. Practically, this means that our ensemble should be considered when execution time is an important metric, given that the predictive performance is comparable to that of the LB ensemble. While these results are promising, it follows that a high reliance on labelled data is prohibitive in a traditional semi-supervised setting and that further analysis is needed. Next, we discuss our results against a real-world intrusion detection database.

4.4   Intrusion Detection Databases

The **ADFA-LD** host-based network intrusion detection database is a publicly available database that contains contemporary attack methodologies and operating systems. ADFA-LD was designed to represent a complete system compromise, from initial penetration through to privilege escalation, thus presenting a realistic scenario for actual hacking as it is conducted in the modern cyberenvironment [4]. It follows that the labels of hack attacks are often lacking, or that attacks need to be labelled on the fly. In this experiment, our aim was to determine how well our LESS-TWE algorithm would perform in such settings.

We consider five different *add new superuser* attacks, which are forms of client-side poisoned executable. Such attacks may lead to exploitation payload, remote operation, staging, system manipulation, data exfiltration, and back-door insertions. The streams are subject to abrupt intrusion attempts, where attacks begin sporadically and dominate the data stream for 250 to 500 instances, before returning to their original frequency.

Next, we study the impact of reducing the number of labels. Table 2 depicts the run times and mean κ values against the ADFA Intrusion data streams. Our results show that there is no significant difference in terms of the κ statistic when comparing LESS-TWE with 90%, 75% and 10% of data having labels on arrival. The algorithm runs very fast, even when faced with a large amount of unlabelled data. We conclude that the LESS-TWE algorithm produces accurate and timely results, even in the presence of 90% data missing labels on arrival. This result is promising in a scenario where the absence of labels should not affect the detection of hack attacks, a concept we plan to study further.

Table 2. Run time (seconds) and κ_m values against ADFA-LD database.

Algorithm	ADFA-SU0		ADFA-SU1		ADFA-SU2		ADFA-SU3		ADFA-SU4	
	Time	κ_m	Time	κ_m	Time	κ_m	Time	κ_m	Time	κ_m
LESS-TWE-90	8.96	0.83	9.05	0.84	8.80	0.83	8.96	0.85	8.64	0.84
LESS-TWE-75	7.82	0.83	8.24	0.84	8.26	0.84	8.26	0.83	0.24	0.84
LESS-TWE-10	13.90	0.84	13.97	0.84	17.90	0.84	15.10	0.84	20.08	0.84

5 Conclusion

This paper focused on learning from evolving streams where the labels may be missing or arriving after a delay. The goal of this study was twofold: firstly, to design fast algorithms to work within the interleaved test-then-train paradigm when there is a potential lack of labels, and secondly, to design algorithms that can detect drifts without relying on the ground truth. Experiments were conducted to evaluate the performance of our LESS-TWE algorithm, while considering the percentage of labelled instances used at each stage of learning. Our results show that our algorithm is very fast and able to produce accurate models with limited labels.

Future work will include exploring the ensemble classifier diversity guarantees by potentially substituting cyclical for stochastic training from novel windows. In this paper, we compared our work to baselines algorithms, as well as the state-of-the-art LB algorithm. Further work will include a comparison with other (ensemble, drift detector) pairs. The reader should note that, in many prior offline semi-supervised learning studies, the amounts of labelled data are typically less than 10%. We plan to extend our research by further investigating these scenarios, focusing on a streaming setting where labels may arrive in bursts. In addition, we plan to further explore additional attack styles as prevalent in the ADFA-LD benchmark database.

References

1. Bifet, A., Gavalda, R.: Learning from time-changing data with adaptive windowing. In: Proceedings of the 2007 SIAM International Conference on Data Mining, pp. 443–448. SIAM (2007)
2. Bifet, A., Holmes, G., Pfahringer, B.: Leveraging bagging for evolving data streams. In: Balcázar, J.L., Bonchi, F., Gionis, A., Sebag, M. (eds.) ECML PKDD 2010. LNCS (LNAI), vol. 6321, pp. 135–150. Springer, Heidelberg (2010). https://doi.org/10.1007/978-3-642-15880-3_15
3. Bifet, A., et al.: New ensemble methods for evolving data streams. In: Proceedings of the 15th ACM SIGKDD International Conference on Knowledge Discovery and Data Mining, pp. 139–148. ACM (2009)
4. Creech, G., Hu, J.: A semantic approach to host-based intrusion detection systems using contiguous and discontinuous system call patterns. IEEE Trans. Comput. **63**, 807–819 (2014)
5. D'Ettorre, S., Viktor, H.L., Paquet, E.: Context-based abrupt change detection and adaptation for categorical data streams. In: Yamamoto, A., Kida, T., Uno, T., Kuboyama, T. (eds.) DS 2017. LNCS (LNAI), vol. 10558, pp. 3–17. Springer, Cham (2017). https://doi.org/10.1007/978-3-319-67786-6_1
6. Flach, P.: Machine Learning: The Art and Science of Algorithms that Make Sense of Data. Cambridge University Press, Cambridge (2012)
7. Gama, J., Medas, P., Castillo, G., Rodrigues, P.: Learning with drift detection. In: Bazzan, A.L.C., Labidi, S. (eds.) SBIA 2004. LNCS (LNAI), vol. 3171, pp. 286–295. Springer, Heidelberg (2004). https://doi.org/10.1007/978-3-540-28645-5_29
8. Haque, A., Khan, L., Baron, M.: Semi supervised adaptive framework for classifying evolving data stream. In: Cao, T., Lim, E.-P., Zhou, Z.-H., Ho, T.-B., Cheung, D., Motoda, H. (eds.) PAKDD 2015. LNCS (LNAI), vol. 9078, pp. 383–394. Springer, Cham (2015). https://doi.org/10.1007/978-3-319-18032-8_30
9. Japkowicz, N., Shah, M.: Evaluating Learning Algorithms: A Classification Perspective. Cambridge University Press, Cambridge (2011)
10. Krawczyk, B., et al.: Ensemble learning for data stream analysis: a survey. Inf. Fusion **37**, 132–156 (2017). ISSN 1566-2535
11. Krempl, G., et al.: Open challenges for data stream mining research. ACM SIGKDD Explor. Newsl. **16**(1), 1–10 (2014)
12. Nishida, K., Yamauchi, K.: Adaptive classifiers-ensemble system for tracking concept drift. In: 2007 International Conference on Machine Learning and Cybernetics, vol. 6, pp. 3607–3612. IEEE (2007)
13. Pesaranghader, A., Viktor, H., Paquet, E.: Reservoir of diverse adaptive learners and stacking fast Hoeffding drift detection methods for evolving data streams. Mach. Learn. **107**(11), 1711–1743 (2018). https://doi.org/10.1007/s10994-018-5719-z
14. Sobolewski, P., Wozniak, M.: Concept drift detection and model selection with simulated recurrence and ensembles of statistical detectors. J. Univ. Comput. Sci. **19**(4), 462–483 (2013)

15. Street, W.N., Kim, Y.S.: A streaming ensemble algorithm (SEA) for large-scale classification. In: Proceedings of the Seventh ACM SIGKDD International Conference on Knowledge Discovery and Data Mining, pp. 377–382. ACM (2001)
16. Zhu, X., Goldberg, A.B.: Introduction to semi-supervised learning. Synth. Lect. Artif. Intell. Mach. Learn. **3**(1), 1–130 (2009)
17. Žliobaitė, I., Bifet, A., Read, J., Pfahringer, B., Holmes, G.: Evaluation methods and decision theory for classification of streaming data with temporal dependence. Mach. Learn. **98**(3), 455–482 (2014). https://doi.org/10.1007/s10994-014-5441-4

Disentangling Aspect and Opinion Words in Sentiment Analysis Using Lifelong PU Learning

Shuai Wang[1(✉)], Mianwei Zhou[2], Sahisnu Mazumder[1], Bing Liu[1],
and Yi Chang[3]

[1] Department of Computer Science, University of Illinois at Chicago, Chicago, USA
shuaiwanghk@gmail.com, sahisnumazumder@gmail.com, liub@cs.uic.edu
[2] Yahoo! Research, Sunnyvale, USA
mianwei@yahoo-inc.com
[3] Artificial Intelligence School, Jilin University, Changchun, China
yichang@acm.org

Abstract. While sentiment analysis can mine valuable information from online reviews, performing a fine-grained sentiment analysis task is very challenging due to the complex patterns in text. In this work, we focus on a Fine-grained Target-based Sentiment Analysis (FTSA) task, which is to identify target-specific aspect words and opinion words. This task is very useful in practice. However, existing solutions cannot generate satisfactory results, especially when we have limited or no labeled data. To provide a holistic solution, we design a novel two-stage approach. Stage one groups the target-related words (call t-words) for a given target, which is relatively easy. Stage two separates the aspect and opinion words from the grouped t-words, which is more challenging due to the lack of sufficient word-level aspect and opinion labels. To address it, we formulate the task in a Positive-Unlabeled (PU) learning setting and incorporate the idea of lifelong learning, which achieves promising results.

1 Introduction

Carrying valuable opinionated information, online reviews have become an important type of big (text) data. The sentiment analysis on them thus plays a crucial role for both customers and manufacturers. Although various sentiment analysis tasks have been studied [13], performing a fine-grained sentiment analysis task remains to be very difficult because of the complex patterns in text, such as sentiment composition, semantic understanding, and word-sense disambiguation. That is probably also why sentiment analysis is still an active research field. In this work, we focus on an important Fine-grained Target-based Sentiment Analysis (FTSA) task.

The FTSA task is defined as: *given a target name, to identify its aspect words and opinion words in a given domain corpus.* Here a target name (or simply

© Springer Nature Switzerland AG 2020
M. Ceci et al. (Eds.): NFMCP 2019, LNAI 11948, pp. 100–115, 2020.
https://doi.org/10.1007/978-3-030-48861-1_7

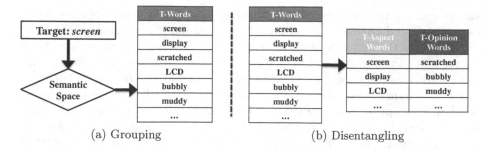

(a) Grouping   (b) Disentangling

Fig. 1. Two-stage approach to Fine-grained Target-based SA (FTSA)

called *target*) can be understood as an aspect category[1], such as *screen, voice,* or *weight.* For example, a customer or manufacturer could be interested in opinions about the target "screen" of a camera (camera is a product or called a **domain**), and wants to find out all related aspect words and opinion words mentioned in the customer reviews. Ideally, one may find aspect words like "LCD," "display", and "resolution," and opinion words like "scratched," "blurry", and "bubbly."

Notice the FTSA problem is very challenging in practice, especially when there is limited or no labeled data. It is also somewhat unrealistic to manually annotate all possible aspect and opinion words for all possible targets in every domain, not to mention that there are always new domains/products coming to the real world. Designing a weakly-supervised or semi-supervised method for this task is thus more practical and useful. To this end, we developed a (semi-supervised) two-stage approach which does not require manual labeling, but only some prior and general knowledge.

Stage one is defined as the process of grouping target-related words. That is, given a target name, we first identify all target-related words (called *t-words*). Note that a *t-word* can be either an aspect word or opinion word. For instance, when the target is *screen,* the t-words "display", "LCD", "scratched", and "bubbly" could be extracted (shown in Fig. 1a). We can achieve this goal by learning the semantic representation of words [2,11]. Specifically, given a target, we extract its semantically similar words from its leaned representation as t-words.

One key issue is that, most semantics learning techniques will inevitably couple the target-related aspect words (called *t-aspect words*) and opinion words (called *t-opinion words*). Here we interpret its cause from a linguistic perspective. Most semantics learning models are developed based on the idea of *distributional hypothesis*: linguistic items occurring with similar contexts have similar meanings [7], so they in fact group two different types of semantic similarity together, namely, conceptual and associative similarity. Conceptual similarity means two words are conceptually similar (likely replaceable), like "dog" and "canine". Associative similarity means two words tend to appear in similar contexts, like "dog" and "bark." The distinction between them is well-known in cognitive science [22] and recently discussed in NLP [11]. In regard to sentiment analy-

---

[1] The terms *target, target name,* and *aspect category* will be used interchangeably.

sis, we can see that t-aspect words "display" and "LCD" and t-opinion words "scratched" and "bubbly" are all mixed based on our given example (Fig. 1a).

In spite of the discussed drawbacks, we argue that the semantics-based models are still quite suitable and helpful for the FTSA task, with the reason being three-fold. First, the mixture benefits the t-words grouping, where both two types of semantic correlation can be jointly extracted. To be concrete, aspect words like "display" could be found because of the conceptual similarity (similar to "screen") and opinion words like "scratched" could also be discovered due to associative similarity (associated with "screen"). Second, many existing or new target extraction techniques can be utilized [13], which paves the way for more accurate results for FTSA (detailed in Sect. 3). Third, those semantics-based models are usually learned in an unsupervised or semi-supervised manner, which meets our learning requirements. However, when we take advantage of those semantics-based models for stage one, we have to overcome their aforementioned drawbacks, which leads to the proposed stage two.

Stage two is defined as: *Given a list of target-related words (t-words), separating them into target-related aspect words (t-target words) and target-specific opinion words (t-opinion words).* Figure 1b shows an example. Notice that the list of t-words is assumed to be given, which is grouped by an existing semantics-based learning technique, so we refer to this problem as *disentangling aspect and opinion words from extracted/grouped target-related words*.

An intuitive solution to this problem is to model it as a word-level binary classification task. That is, to build a classifier to learn and predict t-aspect and t-opinion words. However, this is difficult in practice, because this means that we need both aspect and opinion word-level labels for every domain, which requires intensive human efforts for annotation. Noticing this, we formulate the classification problem in a Positive-Unlabeled (PU) learning setting. The idea is to use general/common opinion words (treating them as positive examples) to distill other opinion words from unlabeled words. However, a notable issue in this PU setting is that the errors from false positive (FP) examples (wrongly predicted opinion words) can be propagated during the PU iterative learning (will be detailed in Sect. 4.3), resulting in more errors and degenerating its performance. To address this issue, we exploit the idea of *lifelong machine learning* [3] and incorporate it into the PU learning process. We name it as Lifelong PU learning (LPU). It works by accumulating the knowledge learned from multiple domains, and uses it to restrict the propagation of FP examples and to ensure the reliability of the newly learned opinion words.

The main contributions of this paper are summarized as: (1) It proposes to perform the FTSA task in a two-stage manner, which does not require manual labeling. (2) It proposes a Lifelong PU (LPU) learning approach to solving the problem of disentangling target-specific aspect and opinion words. To the best of our knowledge, none of the existing studies has employed the LPU technique. (3) Experimental results conducted on multiple real-world review datasets with two target extraction techniques show its effectiveness and extensibility.

2   Related Work

Fine-Grained and Target-Based Sentiment Analysis. Various types of sentiment analysis (SA) research exist in the literature [13,30]. Unlike the coarse-grained SA such as classifying a review document as overall positive or negative [29], fine-grained SA consists of various components such as aspect identification, opinion identification, and polarity classification [25]. In terms of targeted-based SA, most of the existing studies [23,26,27] focused on the target-based polarity classification. However, our work does not lie in this direction. As discussed in Sect. 1, we aim to provide a generic solution for disentangling target-specific aspect and opinion words. In fact, our work can be integrated into other related target-based analysis models and we will show it in our experiments.

Semantic Space and Representation. Semantics-based learning models project words to a semantic space and represent each word as a dense vector. Such semantics-bearing vectors can be created by matrix factorization (e.g. LSI) [5] and topic modeling (e.g., LDA) [2]. Recently, neural word embeddings [17] emerge to learn better semantic representation for words.

Lifelong Machine Learning. Our work is related to lifelong learning (LL) [3]. Regarding sentiment analysis, several LL models have been proposed for improving topic quality [25] and polarity classification [26], but they are not for the FTSA task and not applicable to the word disentangling problem. We also incorporate LL into the PU learning process, which is the first attempt. Although related, our work distinguishes itself from the research field of transfer learning (TL) [16,19], because the LL settings are essentially different from the ones from TL. For example, TL has no knowledge retention. We do not aim to transfer features or use labeled data from a specific source domain to a target domain. Instead, we aim at mining knowledge from previous/seen domains cumulatively, and applying the mined knowledge to a new domain (with imposed constraints). More distinction between LL and TL can be found in a survey book [3].

3   Stage One: Grouping

As discussed in Sect. 1, we group the target-related words (t-words) for a specified target in this stage. The basic idea is to extract its semantically correlated words towards the target in a learned semantic space. Specifically, we use the neural word embedding model [17] to learn word vectors for a given domain corpus, resulting in an embedding matrix $E \in \mathbb{R}^{v \times d}$ where v and d are the size of vocabulary and embedding dimension. Then a semantic similarity matrix $M \in \mathbb{R}^{v \times v}$ is calculated based on the dot product of E and E^T. After that, when a user-specified target is provided, the nearest neighboring words of the target will be returned as t-words, based on their similarity values in M. Notice that other semantics learning models can be used in a similar way [5]. Probabilistic topic models [2] can be used as well, by searching the corresponding topic for the given target and returning the topical words. Notice that this stage is highly similar to the unsupervised target extraction in sentiment analysis [13] and many existing models can be utilized at this stage. The key difference is that, the target name is specified in our setting, so we do not have to perform a full extraction of all possible targets covered by a given corpus, but only focus on the given target (name) by returning its nearest neighbors.

4 Stage Two: Disentangling

4.1 PU Learning Using Word Vectors

This stage separates the given t-words into *t-aspect words* and *t-opinion words*. As discussed in Sect. 1, in order to provide a general approach without manual labeling for any possible domain, we formulate this problem as a binary classification task in a PU learning setting [12]. Clearly, in addition to aspect and opinion words, a domain vocabulary also contains other words like background words. However, as indicated in [18,25], those words do not have a seriously bad effect as they are unlikely to be semantically similar to a given target. Therefore, we assume/treat most of the non-opinion words in the t-words are/as aspect words. This assumption holds well as shown in studies [18,25] and our experiments.

In regard to PU learning, it can be understood as a particular type of semi-supervised learning methods, which learns a binary classifier using only positive and unlabeled examples (with no negative examples). Here P represents a set of data examples with positive labels. In our task, the opinion words from an opinion lexicon will be the words in P, such as "good", "bad" and "angry". In terms of U, it denotes the set of data examples with unknown labels. In our case, other words that are not in the lexicon are in U, where U consists of both (true) opinion words and non-opinion words. With word vectors as features and a set of general opinion words as positive labels, we can build a PU classifier. In our work, we use logistic regression[2] as the PU classifier, which can generate the probabilistic score of a word being in the positive class (i.e., opinion word). In this way, words from U with high prediction scores can be detected as newly identified opinion words, and we can identify more words iteratively with new opinion words being found. However, a notable issue in this PU setting is that the errors from false positive (FP) examples (wrongly predicted opinion words) can be propagated, thus degenerating its performance. To address it, we exploit the idea of *lifelong machine learning* [3] and incorporate it into PU learning. The idea is to exploit the classification knowledge learned from past domains to increase the correctness and reliability of the newly detected opinion words.

---

[2] Other models that provide probabilistic scores can be used as well, e.g., neural networks using Softmax to produce final model outputs, as their prediction of classes would be presented in a distribution normalized to [0,1), i.e., class probability. Here we choose logistic regression (LR) because it can show the improvement achieved by other proposed components more directly (with its simplicity). In our experiments, we use LR as the PU classifier consistently and fairly for each candidate model.

4.2 Lifelong Machine Learning

Lifelong machine learning [3] or *lifelong learning*/LL, works by retaining the knowledge learned from the past tasks and uses it to help future learning, i.e., to help the current or incoming task. It mimics how we humans learn. With regard to sentiment analysis, we (human beings) can learn many opinion expressions in our lives across different domains/products, which enables us to better understand and identify opinion words in a new domain. More details about the LL paradigm can be found in [3]. Following the LL fashion, our system retains the newly learned opinion words every time it has finished processing one domain (one task), treating them as knowledge and accumulating them. The system accumulates such knowledge continuously from continuous domain/product learning. So in any time it has processed N domains and starts to process the $(N+1)$th domain, the accumulated knowledge will be used to help generate more reliable opinion words that are suitable for the $(N+1)$th domain. Based on this general idea, we develop a novel lifelong PU (LPU) learning algorithm.

4.3 Lifelong PU (LPU) Learning

Our proposed LPU algorithm consists of four main steps, namely, knowledge accumulation, current domain setup, knowledge mining and preparation, and restricted PU iterations, which will be detailedly illustrated as follows. The overall algorithm is given in Algorithm 1.

Step 1: Knowledge Accumulation (lines 1–8). This step follows the traditional classification process but with knowledge retention for building a knowledge base from past domains. Specifically, for each domain (task j), we first obtain its vocabulary W_j and semantic representation of words V_j (line 3). With a general opinion lexicon, we then have the lexicon-based opinion words W_j^P, i.e., positive examples, and unlabeled examples W_j^U (line 4). A PU classifier is trained (line 5) and used to predict the probabilistic class scores of words in W_j^U and to find new opinion words W_j^+ (line 6). After that, we retain W_j^+ as knowledge for constructing a knowledge base SKB (line 7).

Step 2: Current Domain Setup (lines 9–13). This step is for the current domain processing setup. The vocabulary words W_i and their semantic representation V_i, lexicon-based opinion words (positive examples) W_i^p and unlabeled words W_i^U of the current domain are first created (lines 10–11). Then we build a hash table H to store the nearest neighbors[3] for all words, which can be easily constructed from the similarity matrix M (see Sect. 3). With the table H established (line 12, and it is a one-time effort), the similarity query becomes a lookup operation. This H not only helps in the current step 2, but also plays a crucial role in the following step 4, as we will see shortly. Based on H, we can find the nearest neighbors for the lexicon-based opinion words and we call them reliable neighbors (line 13). This is an initial constraint, which is also intuitive, as those

---

[3] Simply using top 10 neighbors works consistently well for different domains.

Algorithm 1. Lifelong PU (LPU) Learning

Input: Current domain corpus $D_{i=n+1}$,
    Past domain corpora $\boldsymbol{D}=\{D_1,..,D_j..,D_n\}$
    Opinion words in lexicon W^P, Maximum learning iteration m
    Number of learned words in one iteration l
Output: All newly-extracted opinion words W_i^+ in D_i

1: // Step 1. Knowledge Accumulation
2: **for** each domain corpus $D_j \in \boldsymbol{D}$ **do**
3:   $W_j, V_j \leftarrow GetWordsAndEmbeddings(D_j)$
4:   $W_j^P \leftarrow W_j \cap W^P$, $W_j^U \leftarrow W_j - W_j^P$
5:   $c_j \leftarrow PUClassifier(V_j, W_j^P)$
6:   $W_j^+ \leftarrow OpinionWordPrediction(c_j, W_j^U)$
7:   $\boldsymbol{SKB} \leftarrow \boldsymbol{SKB} \cup W_j^+$ // sentiment knowledge base
8: **end for**
9: // Step 2. Current Domain Setup
10: $W_i, V_i \leftarrow GetWordsAndEmbeddings(D_i)$
11: $W_i^P \leftarrow W_i \cap W^P$, $W_i^U \leftarrow W_i - W_i^P$
12: Create a hash-table H to store top neighbors of all words
13: $W_i^{RN} \leftarrow GetReliableNeighbors(H, W_i^P)$
14: // Step 3. Knowledge Mining and Preparation
15: $W^{SK} \leftarrow FIM(\boldsymbol{SKB})$
16: $W_i^{SK} \leftarrow W_i \cap W^{SK}$ // sentiment-knowledge for domain i
17: $W_i^{RS} = \varnothing$ // reliable learned opinion words
18: $W_i^{PP} = \varnothing$ // current positive prediction (opinion words)
19: $W_i^{NS} = W_i^{RN} \cap W_i^{SK}$ // newly learned sentiment
20: // Step 4. Restricted PU Iterations
21: $t = 0$
22: **while** $t < m$ or W_i^{NS} is not empty **do**
23:   $W_i^{RS} \leftarrow W_i^{RS} \cup W_i^{NS}$ // updating reliable sentiment
24:   $c_i \leftarrow PUClassifier(V_i, W_i^{RS} \cup W_i^P)$
25:   $W_i^{NEW1} \leftarrow MineReliableOpinion(W_i^{SK}, W_i^{PP}, H, l)$
26:   $W_i^{NEW2} \leftarrow ReliableOpinion(W_i^{PP}, W_i^{PP}, H, l)$
27:   $W_i^{NS} \leftarrow W_i^{NEW1} \cup W_i^{NEW2}$
28:   $W_i^{PP} \leftarrow OpinionWordPrediction(c_i, V_i)$
29:   $t = t + 1$
30: **end while**
31: $W_i^+ \leftarrow OpinionWordPrediction(c_i, W_i^U)$

Algorithm 2. $MineReliableOpinion(A, B, H, l)$

1: $S = \varnothing$ // counts positive neighbors for every word in A
2: **for** each a word $w \in A$ **do**
3:   $S \leftarrow countPositiveNeighbors(B, H(w))$
4: **end for**
5: **return** $sortAndReturnTopLWords(A, S, l)$

unlabeled/candidate words which are highly similar to the opinion words known from a lexicon should be more reliable (as opinion words).

Step 3: Knowledge Mining and Preparation (lines 14–19). This step is for mining knowledge and making preparation for later use. With the knowledge accumulated from many past domains and stored in SKB, we can extract the reliable knowledge W^{SK} (line 15). Here we adopt frequent itemset mining (FIM) [1]. The rationale behind it is that: the candidate words frequently predicted as opinion words in many different domains are more trustworthy and confident to be the real opinion words. The intersection of the reliable neighbors W_i^{RN} and reliable knowledge W_i^{SK} initializes W_i^{NS}, the newly learned sentiment (line 19). Lines 16–18 define other variables that are used in step 4, where W_i^{SK} denotes the sentiment knowledge for current domain i, W_i^{RS} the reliable learned sentiment (opinion words) during the PU learning iteration, and W_i^{PP} the newly-predicted opinion words in an ongoing iteration.

Step 4: Restricted PU Iterations (lines 21–31). This step performs iterative PU learning with imposed constraints. As discussed in Sect. 1, in the PU learning setting, the errors from false positive (FP) examples (wrongly predicted opinion words) can be propagated in its iterative learning process (or called self-bootstrapping). For example, if the word "display" is wrongly predicted as opinion word in the first iteration, and when it is added to the positive examples (treated as a newly found positive example) in the second PU learning iteration, it could lead to the mis-classification of more such wrong words, like "screen" and "monitor" which are semantically similar to "display".

The above issue is considered and addressed in this step. Unlike using direct self-bootstrapping methods, here the expansion of the newly-predicted opinion words (as positive examples) in LPU is controlled more strictly. That is, only the reliable ones could be further used. The initialized new opinion words W_i^{NS} have already been restricted (see step 2) and used as initial reliable sentiment W_i^{RS}. During the iterative learning process, it keeps being updated (line 23) by adding only reliable opinion words (line 27).

More specifically, two ways are developed for expanding new reliable opinion words. One way is to learn from the reliable knowledge (line 25) and another way is to learn from its self-predicted results (line 26). Both ways are restricted by the defined reliability score shown in Algorithm 2. This score is calculated based on the number of identified positive neighbors of a candidate word, which is also used for ranking. In Algorithm 2, A denotes the candidate word set and B denotes positive examples. The identified positive neighbors are from the intersection of positive examples (provided by B) and the neighbors of a candidate word (provide by $H(w)$). In each iteration, only the top l ranked words will be trusted and added as new positive examples. When the maximum iteration is met or there are no more new opinion words that the system can learn, the iterative learning process stops and all newly-detected opinion words are returned (line 31).

5   Experiments

5.1   Candidate Methods for Comparison

Adjective Extraction (ADJ): This baseline regards all adjective words as opinion words and others as aspect words. This is a simple but widely used solution. We used POS tagging to extract adjectives. No classifier is trained here.

Part-Of-Speech (POS): The POS features have been reported effective for aspect and opinion extractions. This is also a representative syntax-based app-roach used in many related works [18,25]. Here every word is represented by the POS features of its context, i.e., w_i will be represented as $[POS_{i-1}, POS_i, POS_{i+1}]$. This is used as the word representation for building a classifier.

Latent Semantic Indexing (LSI): LSI is a standard matrix factorization technique to construct latent semantic vectors. Its factorized word-feature corre-lation matrix can be used as word vector representation [21] to build a classifier.

Latent Dirichlet Allocation (LDA): LDA [2] is a classic topic model which discovers hidden topics from documents and groups words into topics. Similar to LSI, the term-topic matrix is used as the word vector representation [14].

Non-Lifelong Learning (NLL): This method is based on our introduced solu-tion but without lifelong learning. It uses the word vectors learned by neural word embeddings to build a classifier.

Lifelong PU (LPU): This is our proposed lifelong PU learning algorithm intro-duced in Algorithm 1.

Lifelong PU minor (LPU-): This is an LPU variant that does not make self-prediction explorations and relies more on the past mined knowledge. In other words, it considers the first type of reliable sentiment only (lines 25 in Algorithm 1). This can be viewed as a conservative version of LPU.

5.2   Experimental Setup

Data. We use a large corpus of Amazon reviews from 20 different domains pro-vided by [15] and the full list is shown in Table 1. For training all PU classifiers, a general opinion lexicon [9] is used so the words appear in it are automat-ically labeled as P. For testing/evaluation, we manually label the aspect and opinion words. Note that here we only manually label those words for the evalu-ation purpose, and we never and will not use those label words for any training. Specifically, three domains from different products are selected, namely, *cell-phone*, *beauty* and *office*. These three domains are also quite different, which help to test the extensibility of the candidate methods. For each domain, three different targets are specified for evaluation. More details are given in Table 1.

Table 1. Detailed information about the domains for evaluation and the full list.

Dataset	#Reviews	#Words	Words in lexicon	Target for evaluation
CellPhone	194,439	28,942	2,764	display, volumes, weight
Beauty	198,502	29,695	2,778	cleansers, fragrance, groomers
Office	53, 258	20,858	2,332	papers, clips, chairs
Full domains	apps for android, amazon instance video, automotive, baby, grocery, health, kindle, tools/home improvement, home and kitchen			

Parameters and Settings. For every candidate method except ADJ, their word vectors/features are learned and used for classification. Specifically, for LSI and LDA, we obtained the term-feature matrix and term-topic matrix. For NLL, LPU- and LPU, we used the skip-gram model [17]. The vector dimension is set to 200 as default and we maintain the same size for LDA and LSI. Logistic regression is used as the classifier for all methods. For LPU, we treat other 19 domains besides the current domain as the past domains to mine knowledge. Notice that for a current domain, only its domain data and the automatically accumulated knowledge will be used, and no other extra domain data will be available, which follows the lifelong learning experimental setting from existing works [3,25]. We empirically set the minimum support to 5 for frequent opinion word mining. We set the maximum iterations m to 10 and the number of words l to learn in each iteration to 50. They work consistently well on different domains/datasets.

5.3 Quantitative Evaluation

Accuracy is used as the metric for evaluation because our task is formulated as a binary classification problem. We also observed that the opinion words

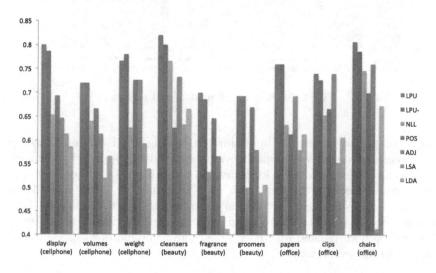

Fig. 2. Acc@150 for all models and targets

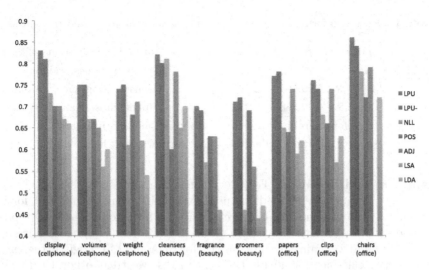

Fig. 3. Acc@100 for all models and targets

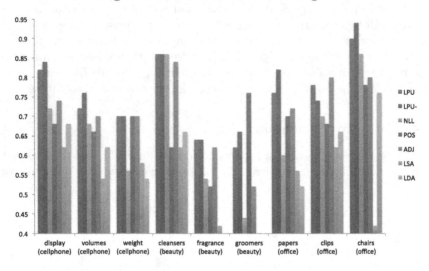

Fig. 4. Acc@50 for all models and targets

(treated as positive example in evaluation/testing) and non-opinion/aspect words (treated as negative example in evaluation/testing) are nearly balanced, so accuracy can reflect the overall identification of both aspect and opinion words.

However, since it is hard to know the exact number of all related words (t-words) to a given target, we use the accuracy@n (acc@n) as our evaluation measure, where n is set to 50, 100, and 150. Specifically, given a target, we first collect its top t-words, and manually label them as opinion or aspect words (as ground truth for evaluation).

We then apply every trained candidate model to classify those top n words to calculate its corresponding acc@n. The results are reported in Fig. 2, 3 and 4. Based on them, we have the following observations:

(1) LPU and LPU- outperform other baselines markedly. LPU improves the best baseline results by 8.29%, 7.11% and 4.00% in acc@150, acc@100, and acc@50. Likewise, LPU- improves the best baseline results by 7.55%, 6.44% and 5.77% in acc@150, acc@100, and acc@50. They demonstrate the effectiveness of lifelong learning.

(2) LPU achieves better performance than LPU- in acc@150, acc@100 but is inferior to LPU- in acc@50. This indicates that LPU is more accurate by considering a big n (more t-words), but LPU- could be more suitable if we only focus on the top-ranked words.

(3) Among other baselines, we observe that NLL and POS perform the best. While POS explicitly reflects the contextual syntax, it is worth noting that the word embeddings used in NLL is also implicitly learned from word-context matrix [11], which implies the syntactic information is very useful in this task.

Table 2. Results for target *volumes* in domain cellphone. Incorrect sentiment words are italicized and marked in red.

Target: Volumes (Domain: CellPhone)		
Model	Aspect	Opinion
LPU	volumes, bass, *undistorted*, volume, Gaga, pitches, sound, *harshness*, cymbals, treble, eq, conf, Mids, reproduces, LiveAudio, *reproducing*, bitrate, Highs, Destroid, *attenuated*,	muddiness, shrill, trebles, distortion, hissy, loud, sibilance, thumping, thump, highs, soundstage, midrange, loudest, *equalization*, overpowered, tinny, *audiofile*, distorted, muddled, piercing,
LPU-	bass, volume, Gaga, pitches, sound, cymbals, treble, conf, Mids, reproduces, bitrate, Highs, Destroid, sounds, lows, levels, vocals, Treble, mids, Fidelity,	*volumes*, muddiness, undistorted, shrill, trebles, distortion, hissy, loud, sibilance, harshness, thumping, eq, thump, *LiveAudio*, highs, *soundstage*, midrange, reproducing, loudest, attenuated,
NLL	volumes, *muddiness*, bass, *undistorted*, trebles, volume, Gaga, pitches, sound, *harshness*, cymbals, treble, *thumping*, eq, conf, Mids, *reproduces*, *thump*, LiveAudio, highs,	shrill, distortion, hissy, loud, sibilance, midrange, *equalization*, tinny, distorted, piercing, muddy, louder, booming, tinnier, quiet, richness, quieter, hissing, boomy, hiss
POS	volumes, *muddiness*, trebles, Gaga, pitches, *sibilance*, cymbals, Mids, LiveAudio, highs, *reproducing*, Highs, *Destroid*, lows, levels, vocals, audiofile, mids, Fidelity, listeners	*bass*, undistorted, shrill, distortion, hissy, *volume*, loud, *sound*, harshness, *treble*, thumping, *eq*, *conf*, *reproduces*, thump, *soundstage*, midrange, loudest, *bitrate*, attenuated
ADJ	volumes, *muddiness*, bass, trebles, volume, Gaga, pitches, sound, *sibilance*, *harshness*, cymbals, *thumping*, eq, conf, Mids, reproduces, *thump*, LiveAudio, highs, soundstage	undistorted, shrill, distortion, hissy, loud, *treble*, loudest, tinny, distorted, *Treble*, resonant, muddy, listenable, louder, booming, tinnier, quiet, richness, quieter, hissing

5.4   Qualitative Evaluation

This subsection shows some example results in Table 2, about the target *volumes* in the cellphone domain. Since LSI and LDA have much poorer performances than others, we do not include their results here. The represented words are the top predicted aspect words and top predicted opinion words from the t-words of a given target. Incorrect opinion words are italicized and marked in red. As we can see, LPU and LPU- better distinguish aspect and opinion words. For example, the opinion word "muddiness" is extracted by them but not by other models. POS identifies many wrong opinion words like "sound" and "volume". Although ADJ is good at extracting adjective opinion words, it misses other opinion words like "muddiness", "sibilance" and "thumping". NLL also misses many opinion words like ADJ.

Similarly, Table 3 shows the results of the target *chairs* in office. For this target, the opinion words are somewhat harder to detect. For example, NLL can correctly identify opinion words, but it can only identify few of them and many

Table 3. Results for target *chairs* in domain office. Incorrect sentiment words are italicized and marked in red.

Model	Aspect	Opinion
LPU	chairs, chair, headrest, armrests, Miller, Herman, Aeron, Hon, seat, Basyx, *Ignition*, reclining, Vue, Chair, VL105, HON, basyx, seats, VL403, backrest,	upholstered, plush, adjustability, rigidness, creak, *draftsman*, slouching, fart, confortable, thickly, tush, strut, exclude, cushy, divine, ornate, slumping, knee, *cradles*, leans,
LPU-	chairs, chair, headrest, armrests, Miller, Herman, Aeron, Hon, seat, Basyx, *Ignition*, reclining, Vue, Chair, VL105, HON, basyx, seats, VL403, backrest,	upholstered, plush, legged, adjustability, rigidness, creak, *castors*, *draftsman*, slouching, confortable, thickly, hips, tush, *wheelchair*, strut, exclude, cushy, divine, ornate, slumping,
NNL	chairs, chair, headrest, armrests, Miller, Herman, Aeron, Hon, seat, Basyx, *Ignition*, reclining, Vue, Chair, VL105, HON, basyx, seats, VL403, backrest	upholstered, plush, rigidness, creak, tush, strut, cushy, divine, slumping, overweight, cozy, supportive, fanciest, numb, adjustable, creaking, *(not found)*, *(not found)*, *(not found)*, *(not found)*
POS	chairs, headrest, armrests, Miller, Herman, Aeron, Hon, seat, Basyx, *Ignition*, reclining, Vue, Chair, VL105, HON, seats, VL403, *upholstered*, HVL531, lumbar,	*chair*, *basyx*, *backrest*, *reclines*, plush, *hydraulic*, adjustability, rigidness, creak, furniture, *draftsman*, slouching, lumber, breathable, cushion, fart, torso, confortable, outfitted, barefoot
ADJ	chairs, chair, headrest, armrests, Miller, Herman, Aeron, Hon, seat, Basyx, *Ignition*, reclining, Vue, Chair, VL105, HON, basyx, seats, VL403, reclines	upholstered, *backrest*, plush, *hydraulic*, rigidness, creak, breathable, fart, confortable, strut, cushy, carpeted, divine, ornate, slumping, *footrest*, unassembled, *armrest*, overweight, *pneumatic*

Target: Chairs (Domain: Office)

others are missed. POS makes some obvious mistakes like treating "chair" as an opinion word. For ADJ, words like "hydraulic" and "pneumatic" are adjective but just for describing the structure of a chair, which are not opinionated. In contrast, LPU and LPU- are quite stable and can give more correct predictions.

5.5   Further Analysis

In order to further evaluate the generality of our proposed approach, we applied it to another popular target extraction technique: topic modeling [2,24]. Specifically, we run LDA [2] for topic generation and then use our algorithm to separate aspect words and opinion words. It also produces reasonably good results as shown in Table 4. Notations in "-(a, b, c, d)" will be explained below.

We also investigated the effect of alleviating FP error propagation in LPU. We denote three types of iterative-learning models as (a), (b), and (c), and they learned with 10 iterations by considering their newly-identified opinion words as positive examples: (a) LPU, using Algorithm 1; (b) A PU model selecting its predicted positive examples ($prob > 0.5$) in the current iteration as P for the next iteration; (c) A PU model that always combines the newly-predicted positive examples with the initial lexicon-based positive examples, without using the constraints in LPU. We also denote NLL as the model (d), which does not learn iteratively.

Table 4. "-" symbol indicates the models behind it do not identify the word.

Topic: Skin (Domain: Beauty)	
Aspect	Newly-identified opinion
face, skin, use, acne (-b), using, just (-b), wash, feel (-b), make, day, product (-b), lotion	dry (-c,d), rid (-c,d), oily (-c,d), drying (-a,c,d), mild (-c,d), notice (-c,d), huge (-d), new (-c,d), ok (-c,d), younger (-c)
Topic: Headset (Domain: CellPhone)	
Aspect	Newly-identified opinion
headset, sound, quality (-b), bluetooth, adapter, hear (-b), ear, volume, headsets, way (-b)	really (-c,d), long (-a), low (-d), away (-c,d), high (-d), short, quite (-c,d), ok, idea (-c,d), close (-c,d)

We now take a further look at Table 4. The "-" symbol indicates that the models following it do not identify the word. Note that here the opinion words from lexicon P are excluded so we can see how those models perform in classifying the unlabeled words U. We observe: 1). Model (c) misses many interesting opinion words like model (d), which indicates that its positive examples remain very similar during all iterations, i.e., it does not learn many new positive examples; 2). Model (b) mis-classifies many aspect words as opinion words as its FP errors propagate iteratively, i.e., the model is confused by the newly-added false positive examples; 3). Model (a), which is LPU, works robustly well.

6   Conclusion and Future Work

This paper discussed the problem of disentangling t-opinion words and t-aspect words from the grouped t-words for fine-grained target-based sentiment analysis (FTSA). We formulated this problem in a PU learning setting and incorporated the lifelong learning idea to overcome the drawback of error propagation in PU learning during the iterative learning process, so as to find more accurate words. To achieve this, a novel lifelong PU learning (LPU) model was proposed. Our experimental results using real-world data demonstrated its effectiveness qualitatively and quantitatively.

Our approach also provides an easily extensible framework for future work, as many components can be further developed. We already discussed some possible directions in the main context of the paper. Here we summarize them and indicate more others. First, one can integrate new semantic representations of words into LPU, such as the contextual word representations ELMo [20] and BERT [6]. Second, as also indicated in Sect. 4, other more advanced or sophisticated models that generate probabilistic output can be used to replace the current PU classifier (logistic regression), such as convolutional neural networks (CNN) [10] and Long Short Term Memory Networks (LSTM) [8]. Third, one may also try exploring other semi-supervised learning methods like Cross-View Training [4], improving the knowledge mining and utilization [25], and leveraging more useful information from the document or sentence level [28].

Acknowledgments. This work was partially supported by a grant from the National Science Foundation (NSF IIS 1838770) and a research gift from Northrop Grumman Mission Systems.

References

1. Agrawal, R., Srikant, R., et al.: Fast algorithms for mining association rules. VLDB **1215**, 487–499 (1994)
2. Blei, D.M., Ng, A.Y., Jordan, M.I.: Latent Dirichlet allocation. J. Mach. Learn. Res. **3**(Jan), 993–1022 (2003)
3. Chen, Z., Liu, B.: Lifelong machine learning. Synth. Lect. Artif. Intell. Mach. Learn. **10**(3), 1–145 (2016)
4. Clark, K., Luong, M.T., Manning, C.D., Le, Q.V.: Semi-supervised sequence modeling with cross-view training. arXiv preprint arXiv:1809.08370 (2018)
5. Deerwester, S., Dumais, S.T., Furnas, G.W., Landauer, T.K., Harshman, R.: Indexing by latent semantic analysis. J. Am. Soc. Inf. Sci. **41**(6), 391 (1990)
6. Devlin, J., Chang, M.W., Lee, K., Toutanova, K.: BERT: pre-training of deep bidirectional transformers for language understanding. In: NAACL (2019)
7. Harris, Z.S.: Distributional structure. Word **10**(2–3), 146–162 (1954)
8. Hochreiter, S., Schmidhuber, J.: Long short-term memory. Neural Comput. **9**(8), 1735–1780 (1997)
9. Hu, M., Liu, B.: Mining and summarizing customer reviews. In: KDD, pp. 168–177. ACM (2004)

10. Kim, Y.: Convolutional neural networks for sentence classification. In: EMNLP (2014)
11. Levy, O., Goldberg, Y., Dagan, I.: Improving distributional similarity with lessons learned from word embeddings. TACL **3**, 211–225 (2015)
12. Li, X.-L., Liu, B.: Learning from positive and unlabeled examples with different data distributions. In: Gama, J., Camacho, R., Brazdil, P.B., Jorge, A.M., Torgo, L. (eds.) ECML 2005. LNCS (LNAI), vol. 3720, pp. 218–229. Springer, Heidelberg (2005). https://doi.org/10.1007/11564096_24
13. Liu, B.: Sentiment analysis and opinion mining. Synth. Lect. Hum. Lang. Technol. **5**(1), 1–167 (2012)
14. Maas, A.L., Daly, R.E., Pham, P.T., Huang, D., Ng, A.Y., Potts, C.: Learning word vectors for sentiment analysis. In: ACL, pp. 142–150 (2011)
15. McAuley, J., Pandey, R., Leskovec, J.: Inferring networks of substitutable and complementary products. In: KDD, pp. 785–794. ACM (2015)
16. Mignone, P., Pio, G.: Positive unlabeled link prediction via transfer learning for gene network reconstruction. In: Ceci, M., Japkowicz, N., Liu, J., Papadopoulos, G.A., Raś, Z.W. (eds.) ISMIS 2018. LNCS (LNAI), vol. 11177, pp. 13–23. Springer, Cham (2018). https://doi.org/10.1007/978-3-030-01851-1_2
17. Mikolov, T., Dean, J.: Distributed representations of words and phrases and their compositionality. In: NIPS (2013)
18. Mukherjee, A., Liu, B.: Aspect extraction through semi-supervised modeling. In: ACL, pp. 339–348. ACM (2012)
19. Pan, S.J., Yang, Q.: A survey on transfer learning. IEEE Trans. Knowl. Data Eng. **22**(10), 1345–1359 (2009)
20. Peters, M.E., et al.: Deep contextualized word representations. In: NAACL (2018)
21. Pu, X., Jin, R., Wu, G., Han, D., Xue, G.R.: Topic modeling in semantic space with keywords. In: CIKM, pp. 1141–1150. ACM (2015)
22. Tversky, A.: Features of similarity. Psychol. Rev. **84**(4), 327 (1977)
23. Vo, D.T., Zhang, Y.: Target-dependent twitter sentiment classification with rich automatic features. In: IJCAI, pp. 1347–1353 (2015)
24. Wang, S., Chen, Z., Fei, G., Liu, B., Emery, S.: Targeted topic modeling for focused analysis. In: KDD, pp. 1235–1244 (2016)
25. Wang, S., Chen, Z., Liu, B.: Mining aspect-specific opinion using a holistic lifelong topic model. In: WWW, pp. 167–176. WWW (2016)
26. Wang, S., Lv, G., Mazumder, S., Fei, G., Liu, B.: Lifelong learning memory networks for aspect sentiment classification. Big Data **2018**, 861–870 (2018)
27. Wang, S., Mazumder, S., Liu, B., Zhou, M., Chang, Y.: Target-sensitive memory networks for aspect sentiment classification. In: ACL, pp. 957–967 (2018)
28. Wang, S., Zhou, M., Fei, G., Chang, Y., Liu, B.: Contextual and position-aware factorization machines for sentiment classification. arXiv preprint arXiv:1801.06172 (2018)
29. Yang, Z., Yang, D., Dyer, C., He, X., Smola, A., Hovy, E.: Hierarchical attention networks for document classification. In: NAACL, pp. 1480–1489 (2016)
30. Zhang, L., Wang, S., Liu, B.: Deep learning for sentiment analysis: a survey. Wiley Interdisc. Rev.: Data Min. Knowl. Discov. **8**(4), e1253 (2018)

Applications

Customer Purchase Behavior Prediction in E-commerce: A Conceptual Framework and Research Agenda

Douglas Cirqueira[1]([✉]) [iD], Markus Hofer[2], Dietmar Nedbal[3] [iD],
Markus Helfert[4] [iD], and Marija Bezbradica[1] [iD]

[1] Dublin City University, Dublin, Ireland
douglas.darochacirqueira2@mail.dcu.ie
[2] Raiffeisenlandesbank Oberösterreich, Linz, Austria
[3] University of Applied Sciences Upper Austria, Steyr, Austria
[4] Maynooth University, Maynooth, Ireland

Abstract. Digital retailers are experiencing an increasing number of transactions coming from their consumers online, a consequence of the convenience in buying goods via E-commerce platforms. Such interactions compose complex behavioral patterns which can be analyzed through predictive analytics to enable businesses to understand consumer needs. In this abundance of big data and possible tools to analyze them, a systematic review of the literature is missing. Therefore, this paper presents a systematic literature review of recent research dealing with customer purchase prediction in the E-commerce context. The main contributions are a novel analytical framework and a research agenda in the field. The framework reveals three main tasks in this review, namely, the prediction of customer intents, buying sessions, and purchase decisions. Those are followed by their employed predictive methodologies and are analyzed from three perspectives. Finally, the research agenda provides major existing issues for further research in the field of purchase behavior prediction online.

Keywords: Consumer behavior · Purchase prediction · Behavior analytics · Machine learning · Data mining · E-commerce · Digital retail

1 Introduction

Daily online activities generate plenty of opportunities for businesses to understand their consumer behavior in E-commerce platforms [1]. Indeed, consumers around the globe purchased $2.86 trillion on the web in 2018, which represented an 18% growth[1]

---

[1] Digital Commerce 360, Global E-commerce Sales 2019. https://www.digitalcommerce360.com/article/global-ecommerce-sales/.

This research was supported by the European Union Horizon 2020 research and innovation programme under the Marie Sklodowska-Curie grant agreement No. 765395; the industry partner Raiffeisenlandesbank Oberösterreich AG; and supported, in part, by Science Foundation Ireland grant 13/RC/2094.

M. Ceci et al. (Eds.): NFMCP 2019, LNAI 11948, pp. 119–136, 2020.
https://doi.org/10.1007/978-3-030-48861-1_8

in online sales compared to the \$2.43 trillion sold in 2017. According to predictions of the purchasing behavior of customers, companies aim to anticipate their needs and provide personalized services [2, 3].

However, consumer behavior itself is well known as a complex pattern among the data mining community [4]. Aiming to predict the likelihood of such patterns, researchers were applying multiple probabilistic and machine learning (ML) statistical models to historical online customer's data, resulting in somewhat reliable probabilities to predict the next customer's steps [5, 6]. That has also increased the complexity of analyzing this literature, given the multiple approaches and datasets available. Previous reviews and surveys related to this topic have usually focused on the specific literature of recommendation systems [7–10]. On the other hand, our focus is on reducing complexity for understanding the step before recommendations, which is the prediction of customer's next purchases, and in visualizing research opportunities in the field.

Therefore, these paper contributions are a novel conceptual framework for analysis and a research agenda. The framework systematically maps this literature regarding datasets adopted, predictive methods, and tasks with their applications. Specifically, the framework reveals three main tasks, namely, prediction of buying sessions, purchase decisions, and customer intents. Next, it provides eight applications enabled by each task. Finally, it illustrates three perspectives on predictive methodologies, and a research agenda with future work opportunities in the field.

The rest of this paper is organized as follows: Sect. 2 describes the research methodology of the literature review; Sect. 3 presents results and the main contributions, followed by final remarks in Sect. 4.

2   Research Methodology

To provide the framework and research agenda proposed, we performed a literature review following systematic guidelines from Watson (2002) [11] and Kitchenham et al. (2009) [12]. Inspired by [13], two research questions and a search query were developed to collect comprehensive literature within the research scope of purchase prediction in E-commerce. Then, the search query was applied in the following scientific databases, well known for containing literature in the field of behavior analytics: Scopus, Web of Science, Science Direct, EBSCO Host (Business Source Complete and Academic Search Complete), Emerald, IEEE Xplore, Association of Information Systems (AIS) library and ACM Digital Library.

– **Search Query**: "(consumer or customer) AND (purchas* OR buy* OR sale* OR shop* OR behavi*) AND (predict* OR forecast*)"

The searches were performed in the abstract field, except for the Web of Knowledge (abstract title and keywords were used) and AIS libraries (full text was used), due to the characteristics of their search engines. The search period has covered papers from

2014 to 2019, only in the English language, which has provided a total of 9824 exported proposals. The next step removed duplicates and had an inclusion filter only to retrieve papers focused on the problem of consumer purchase behavior prediction. That has provided a total of 429 papers.

Next, the exclusion criteria were applied to remove papers not focused on the E-commerce context. At this stage, the total of papers kept was 35. Based on those proposals, backward and forward searches were conducted via Google Scholar, adding 18 and 10 studies, respectively. The final number of papers for extraction and mapping steps was 63. All those results are available at a Github repository (https://github.com/dougcirqueira/literature-review-purchase-prediction).

3   Results

Tables 1 and 2 provide non-exhaustive lists of the proposals selected for this literature review. Table 1 brings single task proposals (prediction of one outcome), while Table 2 provides multi-task proposals (prediction of multiple outcomes).

A Conceptual Framework of Analysis for Customer Purchase Prediction in E-commerce A conceptual framework of analysis aims to optimize the understanding of a complex topic by breaking it down into smaller and comprehensive components [48]. We adopted a systematic literature review approach to developing the conceptual framework of analysis proposed and illustrated in Fig. 1.

The framework has six components. Component 1 defines the dataset types adopted in this literature. Component 2 classifies in dimensions the input data present in those datasets. Component 3 shows the methodologies adopted for constructing features out of the input data, illustrating how consumer behavior is modeled to predictive analytics. Component 4 introduces the predictive methods summarized into four categories. Component 5 shows which tasks enable what applications from component 6, as identified in Subsect. 3.1. Details on each component will be given under the research questions developed in the literature review.

The two research questions developed to conduct the systematic literature review were the guidance for scoping our findings. The results will be presented, reflecting those questions in Subsects. 3.1 and 3.2.

3.1   RQ 1. What Tasks and Applications Have Been Addressed in the Problem of Consumer Purchase Behavior Prediction in E-Commerce?

This research question addresses components 5 and 6 of the proposed framework. It reveals the literature targeting three main tasks within the online purchase prediction problem. Every task has a different prediction outcome, described as follows:

– **Predict Customer Intent (PCI):** Predict the intention of customer visits online. Examples of intention types reported in the literature are purchase oriented or

Table 1. Selected proposals in single task settings (A: Aggregation; R: Rule; P: Personalized Function; L: Learning; CDM: Classical Data Mining; PC: Probabilistic Classifier; DLC: Deep Learning Classifier; CF: Collaborative Filtering)

Ref	Task	Focused applications	Feature construction				Predictive method	Contribution and targeted research gap
			A	R	P	L		
[14]	Buying session (PBS)	B, C, D, F	x				CDM	Real-time predictions in single visits
[15]			x					Feature engineering for clickstream
[16]			x	x				Association rules for fast predictions
[17]			x		x			Feature engineering for popular products
[18]			x	x	x			Feature engineering from customer search
[19]			x					Benchmark over multiple online shops
[20]			x	x				Feature engineering for multiple products
[21]			x	x				Feature engineering with graph metrics
[22]			x	x				K-Nearest Neighbor for fast predictions
[23]			x		x			Feature engineering with motifs in single sessions
[24]			x	x			PC	Prediction over multiple online visits
[25]			x			x	DLC	Feature learning for automatic feature construction
[26]			x	x		x		Real-time predictions in single visits

(*continued*)

Table 1. (*continued*)

Ref	Task	Focused applications	Feature construction				Predictive method	Contribution and targeted research gap
			A	R	P	L		
[27]	Product (PPD)	A, B, C, E, F, G	x	x			CDM	Handle of data unbalancing
[28]			x					Feature engineering with recency and frequency of page views
[29]			x	x			CF	Combination of features from clickstream and transactions for collaborative filtering
[30]			x					Feature engineering for clickstream
[31]			x					Combination of features from clickstream and transactions for collaborative filtering
[32]			x					Feature engineering with product heterogeneity for collaborative filtering
[33]			x	x		x	DLC	Real-time predictions with ensemble and deep learning
[34]						x		Recommendation of bundles of products considering quality and diversity criteria

(*continued*)

Table 1. (*continued*)

Ref	Task	Focused applications	Feature construction				Predictive method	Contribution and targeted research gap
			A	R	P	L		
[35]	Purchase intent (PCI)	B, D, H	x		x		CDM	Feature engineering for clicks diversity
[36]			x					Predict the intensity of user intent
[37]			x	x				Predict intent before an online visit
[38]			x		x	x		Predict intent before/during the online visit

general [35], browsing, searching, purchasing, and bouncing [37]. This task is essential for identifying similar groups of customers, and for applications in which customer segmentation is needed.

- **Predict Buying Session (PBS):** Predict if a current user online session will end up with a purchase or not. This task is interesting for applications that need to capture the general likelihood of the user conversion during his visit online, without details regarding preferences for specific products.
- **Predict Purchase Decisions (PPD):** Predict customers purchase behavior concerning their buying decisions. For instance, to foresee what product or category a customer will buy; to predict the time or period likely to witness a purchase; to predict the next amount customers are likely to spend in their purchases. PPD is the most complex task, as the aim is to predict fine-grained decisions. That is the ideal task for recommending specific products or services to customers.

Those three identified tasks enable a variety of business intelligence applications for online retailers, such as: **A)** Product Recommendations [29]; **B)** Targeted Marketing [16, 42]; **C)** Layout Personalization of E-commerce Landing Pages [17]; **D)** Load balance Optimization to Prioritize Quality of Service for Likely Buyers [14]; **E)** Stock Management Optimization of Products [28, 32]; **F)** Real-time Customer Service [49]; **G)** Purchase Trends Discovery [15]; **H)** Offers Awareness Based on the Detected Intention of Consumers [35].

Table 2. Selected proposals in multi-task settings (A: Aggregation; R: Rule; P: Personalized Function; L: Learning; CDM: Classical Data Mining; PC: Probabilistic Classifier; DLC: Deep Learning Classifier; CF: Collaborative Filtering)

Ref	Task	Focused applications	Feature construction				Predictive method	Contribution and targeted research gap
			A	R	P	L		
[39]	(PBS & PPD)	A, B	x				CDM	Ensemble learning for buying session and product prediction
[40]			x			x	DLC	Feature learning for buying session and product prediction
[41]			x		x	x		Feature learning for buying session and product prediction
[42]	(PPD)	A, E	x				CDM	Feature engineering for product and customer interdependency
[43]			x	x				Feature engineering and ensemble learning for product and time prediction
[44]	(PPD)	E	x		x		CDM	Prediction of repeated buying patterns over multiple sessions
[45]		E				x	DLC	Prediction of repeated buying patterns over multiple sessions
[46]	(PPD)	B, E	x				PC	Feature engineering from purchase emails to predict next time and amount
[47]	(PBS & PPD)	A, B, E	x				PC	Feature engineering for predicting buying session and next amount

3.2 RQ 2. What Methodologies Have Been Adopted to Predict Consumer Purchase Behavior Online?

This research question addresses the components from 1 to 4 of the framework proposed. It provides three perspectives in the predictive methodologies adopted in this literature.

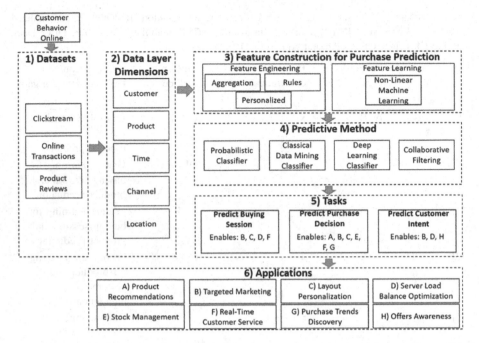

Fig. 1. A conceptual framework of analysis for the literature in behavior and predictive analytics for customer purchase prediction online. (Legends for applications enabled by tasks: A = Product Recommendations; B = Targeted Marketing; C = Layout Personalization; D = Server Load Balance Optimization; E = Stock Management; F = Real-time Customer Service; G = Purchase Trends Discovery; H = Offers Awareness)

Table 3. Dataset types identified in the literature

Dataset	Description	Data layer dimensions involved
Clickstream	Sequences of clicks performed by consumers during their online visits	Customer
Transactions	Purchases executed by customers within the E-commerce	Customer, Product, Time, Channel, Location
Reviews	Text and rating reviews given by customers to specific products	Products

Online Customer Behavior Datasets and their Features. Customer behavior in E-commerce is captured through datasets of past online sessions and shopping logs, which are described in Table 3:

The input data is further classified in the data layer, inspired by [2], in dimensions, which have specific input data features. Every dimension and its features support in explaining and predicting customer behavior from different perspectives, which bring some benefits for predictive tasks on that data, as illustrated in Table 4.

Table 4. Classification of E-commerce data in dimensions and its benefits

E-commerce data dimensions	Input data features $feat_{in}$	Description
Customer	• Demographics • Clicks • Session Variables	Reveals the profile of every consumer, and enables their segmentation. Benefit: tackling the cold start problem
Product	• Value • Description • Status of availability	Relates to the raw characteristics of products online. Benefit: supports the detection of preferences according to product characteristics
Time	• Timestamp • Season	Timestamps of consumer transactions. Benefit: supports the prediction of when events can happen based on previous timestamps and seasonal patterns
Channel	• Customer device • Visit Source	Characteristics of touchpoints between consumers and an E-commerce platform. Benefit: assessment of influences on customer purchase likelihood in different channels
Location	• Neighborhood • City	Information on location of consumers. Benefits: helps in identifying patterns according to the spatial placement of consumers [43, 46]

Feature Construction for Purchase Prediction. In this Subsection, we use a formal notation to explain the feature construction process. The input data features $feat_{in}$ described previously serve as the basis for feature construction, from which is derived new descriptive features $feat_{eng\_out}$ to capture historical patterns, which can indicate the probability of purchase. Two methodologies are adopted to create descriptive features. The first is Feature Engineering, where domain expertise is used to think of a function or rule f_{eng} to apply on input data features $feat_{in}$ present in a dataset, which are related to a current customer transaction Ti. This process can be shaped by conditions $cond_n$ to capture relationships between multiple input data features. The Feature Engineering process can be described in Eq. 1.

$$feat_{eng\_out} = f_{eng}(D,\ Ti, feat_{in},\ cond_1,\ cond_n) \tag{1}$$

The second methodology for feature construction is Feature Learning, in which a function f_{learn} to create new features is an unsupervised ML model, which automatically derives new explanatory features. For instance, researchers extract Latent Representations, or hidden layer weights $feat_{learn\_out}$ learned during training time of a Recurrent Neural Network or Autoencoder model, carrying hidden correlations and relationships between variables. This learning process is conditioned by the target outcome $targ_{out}$ and a cost function $cost_f$, which represent the desired outcome of the learned representation, and how the weights of the hidden layer will be learned. The desired outcome is, for instance, a binary label for predicting buying sessions, or a multi-category label for predicting purchase decisions regarding products. The Feature Learning process is described in Eq. 2.

$$feat_{learn\_out} = f_{learn}\left(D, Ti, feat_{in}, targ_{out}, cost_f\right) \tag{2}$$

Table 5 illustrates examples of those methodologies in action.

Table 5. Methodologies for Customer behavior Feature Construction

Methodology	Function	Description	Examples of New Features $feat_{eng\_out}/feat_{learn\_out}$
Feature engineering	Aggregation	**Examples of Functions:** Count, Sum, Timing, Average, Variance, Ratio $f_{eng} = Average\ \frac{\sum_{i=1}^{n} Amount_i}{n}$ $feat_{in} = clicks$ $cond_1 = product\ category$	**Example of Feature: Average.** Number.Clicks.On.Category [27]
	Rules	**Example of Rule:** Is Purchase in a Shopping Holiday? $f_{eng} = binaryrule$ $feat_{in} = time$ $cond_1 = is\ time.weekday\ a\ holiday?$	**Examples of Feature: 1 or 0** indicating if the rule condition is satisfied or not [26]
	Personalized Functions	**Example of Personalized Function:** Entropy to detect how diverse are customer purchases $f_{eng} = Entropy - \sum_{i \in C} p_i \log_2 p_i$ $feat_{in} = page\ views$ $cond_1 = product\ category$	**Examples of Feature: Diversity.**Browsed.Categories [18]
Feature learning	Non-linear machine learning models	**Examples of Functions:** Autoencoders, Recurrent Neural Networks, $f_{eng} = Autoencoder$ $feat_{in} = (feat_{in\_1},...,feat_{in\_m})targ_{out} = (feat_{in\_1},...,feat_{in\_m})$ $cost_f = Mean\ Squared\ Error$	**Examples of Features:** Compressed representation of input data by the weights of a hidden Autoencoder layer $(weight_1,...,weight_{m/x})$

Predictive Methods. Researchers have been working with ML and probabilistic methods to predict the complex customer purchase behavior online [5]. Based on the conceptual framework, we summarize the predictive models adopted into four categories, with their advantages and disadvantages. It is provided examples of particular methods within each category, specifically for purchase prediction in E-commerce. We illustrate in Table 6 how those models compare concerning their characteristics and suitability for tasks identified in Subsect. 3.1.

The characteristics analyzed are a) *Suitability for Real-Time*: concerning usual time required for training, if any, and for providing predictions in production settings; b) *Interpretability*: concerning the capacity of providing explanations for why a predicted outcome is given by the model; c) *Sequential Modeling*: it illustrates if a predictive method is able to model the customer activities sequentially. That is important when researchers want to explicitly analyze the influence of past purchases in current customer actions; d) *Feature Construction Function*: reveals what methodology and function are usually adopted for feature construction when applying the predictive method analyzed.

Table 6. Predictive methodologies

Predictive method	Example	Characteristics of predictive method				Suitability for purchase prediction tasks		
		Suitability for Real-Time	Interpretability	Sequential Modeling	Feature Construction Function	PCI	PBS	PPD
Probabilistic classifier	Bayesian Classifier, Hidden Markov Model	High	High	High	Aggregation & Rules		X	X
Classical Data Mining Classifiers	Unsupervised clustering: K-means	Low	High	Low	Aggregation & Rules & Personalized	X		
	Association rules: apriori algorithm	High	High	Low			X	
	Instance-based: K-nearest neighbor	High	High	Low			X	
	Linear ML: logistic regression, decision Tree	High	High	Low			X	X
	Ensemble Learning:\ XGBoost, Adaboost, Majority Voting	Low	Low	Low			X	X
Deep learning classifier	Non-linear ML: RNNs, LSTMs, GRUs	Low	Low	High	Aggregation & Non-Linear Feature Learning		X	X
Collaborative filtering	Matrix factorization and factorization Machines	High	Low	Low	Aggregation & Rules			X

Details regarding each predictive methodology are provided as follows.

- **Probabilistic Classifier:** A model that uses probability theory to model the uncertainty in the data. **Advantage:** Usually, it requires a few numbers of engineered features, which makes them a feasible choice for real-time settings, as well as the natural capacity of sequentially modeling short-term patterns in events. **Disadvantage:** It is difficult to capture the effects of long-term patterns in customer behavior. However, this capacity can be achieved in the cost of increasing model complexity and processing time.
  - *Bayesian Classifier:* Estimates conditional probability distributions based on the influence of given features to output a specific prediction. In [42], authors predict purchase decisions by analyzing the influence of sequential purchases, number, and duration of visits to compute probabilities for the customer choice of a specific product or time of purchase.
  - *Hidden Markov Model:* A generalization of a probabilistic mixture model, where the probability of an event, such as a purchase, depends on the occurrence of hidden variables through a sequential Markov process modeling a previous customer action [24].
- **Classical Data Mining Classifiers:** Those models work by learning similarities between feature vectors of buying sessions, intents, and purchase decisions. **Advantage:** Most of the approaches in this category perform well even with small or medium dataset sizes, which makes some of them suitable for real-time settings. **Disadvantage:** Authors adopting this methodology usually need to perform extensive Feature Engineering to achieve good prediction results, also for detecting sequential patterns.
  - *Unsupervised Clustering:* Unlabeled sessions and purchase transactions are input to a model which will discover patterns in similar instances and group them for providing predictions. For example, [37, 38] adopt the K-means algorithm to segment customers based on variables regarding their clickstream behavior.
  - *Association Rules:* Enables the discovery of associations between features, which can reveal rules with high confidence to indicate probabilities of sessions ending up with a purchase [16].
  - *Instance-Based:* Model which classify new data instances based on similar cases and their features. In [22], authors employ K-Nearest Neighbor to predict buying sessions according to previous examples of sessions, with similar features, which ended up with a purchase.
  - *Linear ML:* Machine learning models which assume a linear decision boundary between buying and non-buying sessions, or feature vectors representing purchase decisions of customers. However, the kernel trick can be adopted to detect non-linear relationships between features [50], or Feature Engineering to create combinations between multiple features [27].
  - *Ensemble Learning:* stacking of various weak predictive models together to build up a robust model for providing predictions [20].

- **Deep Learning Classifiers**: ML models which can naturally learn complex and non-linear decision boundaries and relationships in the dataset. **Advantage**: These models can be powerful in modeling long-term influences of past customer events on current decisions [25], and do not require extensive Feature Engineering, as they have Feature Learning built-in. **Disadvantage**: This method usually requires massive amounts of data, which makes it hard for usage with new customers and a few purchases [40, 41]. The interpretability of predictions is also an issue.
- **Collaborative Filtering:** Classical model applied in recommendation systems. This approach models customers and products in a utility matrix based on their clicks, views, reviews, or purchases, which is then factorized to provide latent factors representing the likelihood of customers choosing similar products [29, 30, 44]. That is the favorite model adopted by researchers focusing on predicting purchase decisions, but it is also utilized in predicting buying sessions [14]. **Advantage**: One of the most flexible approaches for multiple types of features in different E-commerce platforms. It also scales well with more customers and products being added in a dataset. **Disadvantage**: The utility matrix is usually sparse, as most of the customers have not viewed many of the products available in an E-commerce platform. Therefore, it is challenging to predict purchases for new customers, and it is important to think of Feature Engineering for creating features that can overcome such issues.

3.3 State-of-the-Art Performance

To have a fair comparison between the identified predictive methodologies, for every specific task, we grouped the existing proposals by the predictive methodology adopted. We evaluated only the F1 score and Area Under Receiver Operating Characteristic Curve (AUC) reported by those. Our choice for those metrics considers the fact that datasets in this literature are usually unbalanced, with few occurrences of purchases, and it is well known that F1 and AUC scores are the ideal metrics in unbalanced scenarios [51]. Table 7 illustrates the average results obtained from predictive methodologies for suitable tasks where they can be applied. It is not reported performance for predicting customer intent as the authors did not adopt the mentioned metrics.

Classical Data Mining Classifiers are the current state-of-the-art for Predicting Buying Sessions, specifically Ensemble learners [20] and Support Vector Machines [19]. Those are followed by Deep Learning classifiers. It is interesting to observe the drop in performance when going to the task of Predicting Purchase Decision, which proves it is the most complex task due to the fine-grained predictions aimed at it. Concerning performance, the classical Collaborative Filtering approach is the most robust, comprised of a Latent Factor Model [30] and Matrix Factorization [31]. Those are followed by Classical Data Mining and Deep Learning classifiers.

Table 7. State of the Art Results for Predicting Buying Sessions and Purchase Decisions

Predict buying session			
F1		**AUC**	
Classical data mining classifier [20]	97.20%	Classical data mining classifier [19]	84%
Deep learning classifier [26]	87.94%	Deep learning classifier [25]	83.90%
Classical data mining classifier [15]	82.91%	Classical data mining classifier [14]	75%
Predict Purchase Decision			
F1		**AUC**	
Collaborative Filtering [30]	53%	Collaborative Filtering [31]	87.94%
Classical Data Mining Classifier [28]	43.62%	Deep Learning Classifier [41]	86%
Collaborative Filtering [32]	42%	Classical Data Mining Classifier [39]	85%

3.4 Research Agenda

We derive a research agenda based on the targeted research gaps and findings of this review, containing the following directions:

- **Sequential Learning**: Few proposals have explored sequential ML models in this literature. Examples are recurrent neural networks, which are only adopted in three studies [25, 33, 40]. Such models are indicated to learn the evolving consumer behavior over time, and sequential patterns such as "She is buying a phone case after purchasing a smartphone".
- **Interpretability**: It is noticed the majority of authors reporting higher performance as those applying Classical Data Mining and Deep Learning classifiers, which also have a black-box nature. Indeed, interpretability seems not to be the focus of this recent literature.
- **Customer Data and General Data Protection Regulation (GDPR)**: Given the rise of privacy policies with GDPR in Europe, it is needed more research on the trade-off between the amount of data required and protection of customers' privacy, regarding the performance of purchase prediction tasks.
- **Dataset for benchmarking:** There is no clear consensus on datasets for state-of-the-art comparison in this literature, as many studies have used private data. However, we observed a significant adoption of the Recsys 2015 challenge data [17, 25, 31, 39, 40, 42], which suggests this dataset as a candidate in this regard.
- **Evaluation in Multiple E-commerce Platforms:** Most researchers evaluate their proposed predictive methods in a single dataset, or focus on specific E-commerce settings. Therefore it is hard to argue their methodologies are general for multiple E-commerce platforms, such as general-purpose and specialized marketplaces.
- **Feature Engineering and Feature Learning:** It was noticed that the well-performing proposals adopting Classical ML models had been heavily investing in

Feature Engineering. However, more investigation in the field of Feature Learning is recommended in this area, or the combination of those two methodologies in purchase prediction online.

- **Creation Process of Personalized Feature Engineering Functions:** Some researchers explore the creation of personalized functions in Feature Engineering, such as the popularity of a product [17], the diversity of customer behavior [18, 35] and graph metrics [21]. It could be relevant to map this creation process, and help other researchers in establishing such novel features for customer behavior online.
- **A Framework for Purchase Prediction Tasks in E-commerce:** Existing proposals focus on one of the three tasks identified, but there is a lack of a view into how those tasks can work together. Therefore, further research could be taken to provide a framework which aligns the identified tasks in this review.

4   Final Remarks

This study presents a systematic literature review of recent proposals in consumer purchase prediction in E-commerce. A novel conceptual framework provides lenses in the state-of-the-art of this field. It is noticed that, despite the broad literature, there is still a need for an in-depth investigation of specific directions. Therefore, a research agenda is provided, illustrating potential future work demands.

A next step would be to adopt a benchmark dataset, and evaluate predictive methodologies in multi-task settings, such as to forecast the next product, purchase time, or amount a customer will likely buy. Therefore, it is relevant to investigate the construction of a framework for purchase prediction, which considers the combination of three tasks identified in this review.

References

1. Agnihotri, R., Dingus, R., Hu, M.Y., Krush, M.T.: Social media: influencing customer satisfaction in B2B sales. Ind. Mark. Manage. **53**, 172–180 (2016)
2. Bradlow, E.T., Gangwar, M., Kopalle, P., Voleti, S.: The role of big data and predictive analytics in retailing. J. Retail. **93**(1), 79–95 (2017)
3. Le, D.-T., Fang, Y., Lauw, H.W.: Modeling sequential preferences with dynamic user and context factors. In: Frasconi, P., Landwehr, N., Manco, G., Vreeken, J. (eds.) ECML PKDD 2016. LNCS (LNAI), vol. 9852, pp. 145–161. Springer, Cham (2016). https://doi.org/10.1007/978-3-319-46227-1_10
4. Erevelles, S., Fukawa, N., Swayne, L.: Big data consumer analytics and the transformation of marketing. J. Bus. Res. **69**(2), 897–904 (2016)
5. Shmueli, G., et al.: To explain or to predict? Stat. Sci. **25**(3), 289–310 (2010)
6. Martens, D., Provost, F., Clark, J., de Fortuny, E.J.: Mining massive fine-grained behavior data to improve predictive analytics. MIS Q. **40**(4), 869–888 (2016)
7. Ricci, F., Rokach, L., Shapira, B.: Introduction to recommender systems handbook. In: Ricci, F., Rokach, L., Shapira, B., Kantor, P. (eds.) recommender systems handbook, pp. 1–35. Springer, Boston (2011). https://doi.org/10.1007/978-0-387-85820-3_1
8. Bobadilla, J., et al.: Recommender systems survey. Knowl.-Based Syst. **46** 109–132 (2013)

9. Lu, J., et al.: Recommender system application developments: a survey. Decis. Support Syst. **74**, 12–32 (2015)
10. Isinkaye, F.O., Folajimi, Y.O., Ojokoh, B.A.: Recommendation systems: principles, methods and evaluation. Egypt. Inf. J. **16**(3), 261–273 (2015)
11. Webster, J., Watson, R.T.: Analyzing the past to prepare for the future: writing a literature review. MIS Q. **26**, xiii–xxiii (2002)
12. Kitchenham, B., Brereton, O.P., Budgen, D., Turner, M., Bailey, J., Linkman, S.: Systematic literature reviews in software engineering–a systematic literature review. Inf. Softw. Technol. **51**(1), 7–15 (2009)
13. Akter, S., Wamba, S.F.: Big data analytics in e-commerce: a systematic review and agenda for future research. Electron. Mark. **26**(2), 173–194 (2016)
14. Zeng, M., Cao, H., Chen, M., Li, Y.: User behaviour modeling, recommendations, and purchase prediction during shopping festivals. Electron. Mark. **29**(2), 1–12 (2018)
15. Jia, R., Li, R., Yu, M., Wang, S.: E-commerce purchase prediction approach by user behavior data. In: 2017 International Conference on Computer, Information and Telecommunication Systems (CITS), pp. 1–5. IEEE (2017)
16. Suchacka, G., Chodak, G.: Using association rules to assess purchase probability in online stores. Inf. Syst. e-Bus. Manag. **15**(3), 751–780 (2017)
17. Chen, C., Xiao, J., Hou, C., Yuan, X.: Improving purchase behavior prediction with most popular items. IEICE Trans. Inf. Syst. **100**(2), 367–370 (2017)
18. Niu, X., Li, C., Yu, X.: Predictive analytics of e-commerce search behavior for conversion. In: Twenty-Third Americas Conference on Information Systems (2017)
19. Lee, M., Ha, T., Han, J., Rha, J.Y., Kwon, T.T.: Online footsteps to purchase: exploring consumer behaviors on online shopping sites. In: 2015 Proceedings of the ACM Web Science Conference. ACM (2015)
20. Boroujerdi, E.G., et al.: A study on prediction of user's tendency toward purchases in websites based on behavior models. In: 2014 6th Conference on Information and Knowledge Technology (IKT), pp. 61–66. IEEE (2014)
21. Baumann, A., Haupt, J., Gebert, F., Lessmann, S.: Changing perspectives: using graph metrics to predict purchase probabilities. Expert Syst. Appl. **94**, 137–148 (2018)
22. Suchacka, G., Skolimowska-Kulig, M., Potempa, A.: A k-nearest neighbors method for classifying user sessions in e-commerce scenario. J. Telecommun. Inf. Technol. **3**, 64–69 (2015)
23. Lin, W., Milic-Frayling, N., Zhou, K., Ch'ng, E.: Predicting outcomes of active sessions using multi-action motifs. In: IEEE/WIC/ACM International Conference on Web Intelligence, pp. 9–17, October 2019
24. Park, C.H., Park, Y.H.: Investigating purchase conversion by uncovering online visit patterns. Mark. Sci. **35**(6), 894–914 (2016)
25. Sheil, H., Rana, O., Reilly, R.: Predicting purchasing intent: automatic feature learning using recurrent neural networks (2018). arXiv preprint arXiv:1807.08207
26. Sakar, C.O., Polat, S.O., Katircioglu, M., Kastro, Y.: Real-time prediction of online shoppers' purchasing intention using multilayer perceptron and LSTM recurrent neural networks. Neural Comput. Appl. **31**(10), 6893–6908 (2019)
27. Li, Q., Gu, M., Zhou, K., Sun, X.: Multi-classes feature engineering with sliding window for purchase prediction in mobile commerce. In: 2015 IEEE International Conference on Data Mining Workshop (ICDMW), pp. 1048–1054. IEEE (2015)
28. Iwanaga, J., Nishimura, N., Sukegawa, N., Takano, Y.: Estimating product-choice probabilities from recency and frequency of page views. Knowl.-Based Syst. **99**, 157–167 (2016)

29. He, T., Yin, H., Chen, Z., Zhou, X., Luo, B.: Predicting users' purchasing behaviors using their browsing history. In: Sharaf, Mohamed A., Cheema, M.A., Qi, J. (eds.) ADC 2015. LNCS, vol. 9093, pp. 129–141. Springer, Cham (2015). https://doi.org/10.1007/978-3-319-19548-3_11

30. Jia, R., Li, R.: Modeling user purchase preference based on implicit feedback. In: CSCWD, pp. 832–836. IEEE (2018)

31. Park, C., Kim, D., Yang, M.C., Lee, J.T., Yu, H.: Your click knows it: predicting user purchase through improved user-item pairwise relationship (2017). arXiv preprint arXiv: 1706.06716

32. Nishimura, N., Sukegawa, N., Takano, Y., Iwanaga, J.: A latent-class model for estimating product-choice probabilities from clickstream data. Inf. Sci. **429**, 406–420 (2018)

33. Singhal, R., et al.: Fast online 'next best offers' using deep learning. In: Proceedings of the ACM India Joint International Conference on Data Science and Management of Data. CoDS-COMAD 2019, pp. 217–223. ACM, New York (2019)

34. Bai, J., et al.: Personalized bundle list recommendation. In: The World Wide Web Conference. ACM (2019)

35. Zheng, B., Liu, B.: A scalable purchase intention prediction system using extreme gradient boosting machines with browsing content entropy. In: 2018 IEEE International Conference on Consumer Electronics (ICCE), pp. 1–4. IEEE (2018)

36. Minjing, P., Xinglin, L., Ximing, L., Mingliang, Z., Xianyong, Z., Xiangming, D., Mingfen, W.: Recognizing intentions of e-commerce consumers based on ant colony optimization simulation. J. Intell. Fuzzy Syst. **33**(5), 2687–2697 (2017)

37. Schellong, D., Kemper, J., Brettel, M.: Generating consumer insights from big data click-stream information and the link with transaction-related shopping behavior. In: Proceedings of the 25th European Conference on Information Systems (ECIS) (2017)

38. Schellong, D., Kemper, J., Brettel, M.: Clickstream data as a source to uncover consumer shopping types in a large-scale online setting. In: ECIS. Research Paper 1 (2016)

39. Romov, P., Sokolov, E.: Recsys challenge 2015: ensemble learning with categorical features. In: Proceedings of the 2015 International ACM Recommender Systems Challenge, vol. 1. ACM (2015)

40. Wu, Z., Tan, B.H., Duan, R., Liu, Y., Mong Goh, R.S.: Neural modeling of buying behaviour for e-commerce from clicking patterns. In: Proceedings of the 2015 International ACM Recommender Systems Challenge, vol. 12. ACM (2015

41. Vieira, A.: Predicting online user behaviour using deep learning algorithms. arXiv preprint arXiv:1511.06247 (2015)

42. Yeo, J., Kim, S., Koh, E., Hwang, S.w., Lipka, N.: Predicting online purchase conversion for retargeting. In: Proceedings of the Tenth ACM International Conference on Web Search and Data Mining, pp. 591–600. ACM (2017)

43. Li, D., Zhao, G., Wang, Z., Ma, W., Liu, Y.: A method of purchase prediction based on user behavior log. In: 2015 IEEE International Conference on Data Mining Workshop (ICDMW), pp. 1031–1039. IEEE (2015)

44. Liu, G., et al.: Repeat buyer prediction for e-commerce. In: Proceedings of the 22nd ACM SIGKDD International Conference on Knowledge Discovery and Data Mining, pp. 155–164. ACM (2016)

45. Guo, L., Hua, L., Jia, R., Zhao, B., Wang, X., Cui, B.: Buying or browsing?: predicting real-time purchasing intent using attention-based deep network with multiple behavior. In: Proceedings of the 25th ACM SIGKDD International Conference on Knowledge Discovery & Data Mining, pp. 1984–1992, July 2019

46. Kooti, F., Lerman, K., Aiello, L.M., Grbovic, M., Djuric, N., Radosavljevic, V.: Portrait of an online shopper: understanding and predicting consumer behavior. In: Proceedings of the

Ninth ACM International Conference on Web Search and Data Mining, pp. 205–214. ACM (2016)
47. Panagiotelis, A., Smith, M.S., Danaher, P.J.: From amazon to apple: modeling online retail sales, purchase incidence, and visit behavior. J. Bus. Econ. Stat. **32**(1), 14–29 (2014)
48. Green, H.E.: Use of theoretical and conceptual frameworks in qualitative research. Nurse Res. **21**, 6 (2014)
49. Tang, L., Wang, A., Xu, Z., Li, J.: Online-purchasing behavior forecasting with a firefly algorithm-based SVM model considering shopping cart use. Eurasia J. Math. Sci. Technol. Educ. **13**(12), 7967–7983 (2017)
50. Schölkopf, B.: The kernel trick for distances. In Advances in Neural Information Processing Systems, pp. 301–307 (2001)
51. Jeni, L.A., Cohn, J.F., De La Torre, F.: Facing imbalanced data–recommendations for the use of performance metrics. In 2013 Humaine Association Conference on Affective Computing and Intelligent Interaction, pp. 245–251. IEEE, September 2013

Hough Transform as a Tool for the Classification of Vehicle Speed Changes in On-Road Audio Recordings

Elżbieta Kubera[1]([⊠]) [iD], Alicja Wieczorkowska[2][iD], and Andrzej Kuranc[1][iD]

[1] University of Life Sciences in Lublin, Akademicka 13, 20-950 Lublin, Poland
{elzbieta.kubera,andrzej.kuranc}@up.lublin.pl
[2] Polish-Japanese Academy of Information Technology,
Koszykowa 86, 02-008 Warsaw, Poland
alicja@poljap.edu.pl

Abstract. Spectrogram is a very useful sound representation, showing frequency contents as a function of time. However, the spectrogram data are very complex, as they may contain both lines or curves corresponding to partials (harmonic or not), whose frequency changes in time, as well as noises of various origin. In this paper, we address the extraction of line parameters from spectrograms for audio data, recorded for cars passing by an audio recorder. These lines represent pitched sounds, and the frequency along these lines is usually related to the vehicle speed. Our goal is to detect whether the vehicle is slowing down, speeding, or maintaining approximately constant speed. However, the lines may be broken, they bent when the car is passing the microphone because of the Doppler effect, which is strongest when very close to the microphone, and they are on the noisy background. Our goal was to elaborate a methodology, which extracts a simple representation of parameters of these lines (possibly broken, curvy and in noise), and allows detecting the behavior of drivers when passing the measurements point, e.g. near the radar. Audio data can be very useful here, as they can be recorded at low visibility. The proposed methodology, together with the results for on-road recorded audio data, are presented in this paper. This methodology can be then applied in works on road safety issues.

Keywords: Speed changes detection · Hough transform · Audio signal analysis

1 Introduction

Road accidents in majority of cases are caused by a failure to yield the right-of-way, or by excessive speed, inadequate to the given road conditions. This information is confirmed by numerous, detailed studies on road incidents and their consequences [1–3]. It is also important that the values of vehicle speed, as circumstances of accident, vary in a wide range, from relatively low speeds in urban areas to high speeds on expressways and motorways.

© Springer Nature Switzerland AG 2020
M. Ceci et al. (Eds.): NFMCP 2019, LNAI 11948, pp. 137–154, 2020.
https://doi.org/10.1007/978-3-030-48861-1_9

Personal features of a driver and his or her habits affect the reactions in dangerous driving situations [4,5]. Usually, drivers are classified according to the level of their "aggressiveness" in driving [5–8]. An aggressive driver is characterized by high speeds of driving and numerous and sudden changes of instantaneous speed, which are associated with periods of acceleration and braking.

The higher speed variations, the greater the interactions between the vehicles on the road and the higher the associated danger [9]. Besides aggressive drivers, careful drivers can also be identified. They try to maintain a constant moderate speed and avoid rapid acceleration and braking, which together are indicators of safe behavior. Many drivers are aware of the impact of speed on the road accidents occurrence. However, they believe that road accidents are caused not only by driving too fast but also by driving too slow, which implicates dangerous behavior of other drivers [4,10].

Economic development is associated with an increase in the number of road transport means. This is followed by the development of infrastructure, and it also requires the introduction of traffic monitoring and control systems. The systems dedicated for speed measurements and vehicle classification contribute to the road safety and traffic fluency. Many transport agencies often use the results of speed tests as the basis of decisions on setting speed limits, traffic signs, synchronizing traffic lights, and assessing their effectiveness [11].

Numerous study works clearly indicate the possibilities of improving traffic safety through comprehensive implementation of traffic management and vehicle speed management systems [12–14]. They can be based on magnetic induction, piezoelectric effect, Doppler effect and computer video analysis techniques.

Traffic measurement technologies can be classified into intrusive and non-intrusive methods [15]. The technologies of the first group basically consist in placing a recorder and a sensor on or in the road:

- Pneumatic road tubes, placed across the road lanes to detect vehicles by means of pressure changes that are generated by a vehicle tyre passing over.
- Piezoelectric sensors: the sensors are placed in a groove along the roadway surface of the lane(s) monitored.
- Inductive loops: the loops are embedded into roadways; they generate a magnetic field [16].

Non-intrusive consist in remote observations:

- Manual counts: trained observers gather traffic data, e.g. vehicle occupancy rate, pedestrians and vehicle classifications.
- Passive and active infra-red sensors: the presence, speed and type of vehicles are detected based on the infrared energy radiating from the detection area.
- Passive magnetic sensors, fixed under or on top of the roadbed.
- Microwave radar: this technology can detect moving vehicles and their speed (Doppler radar) [17].
- Ultrasonic and passive acoustic methods: the devices emit sound waves to detect vehicles by measuring the time needed for the signal to return to the device. The passive acoustic sensors are placed alongside the road and can collect vehicle counts, speed and classification data [18].

– Video image detection, gaining popularity recently: video cameras record registration plates, vehicle type and speed [19–21].

These technologies differ in their installation costs; they have advantages and disadvantages [15,22–24]. Almost all of them allow measuring vehicle speed, but acceleration measurements are not taken into account. Therefore, the development of such systems is needed.

There are studies on the determination of the speeds of vehicles using acoustic waves generated by passing vehicles [25,26], or using an on-board microphone [27]. In particular, an acoustic vector sensor and sound intensity measurement techniques are applied in these methods [28]. They utilize sophisticated algorithms for the sound intensity processing in the domain of time and frequency. The obtained results indicate the potential of these methods, and a possibility of using them as a supplementation of currently employed techniques in measurements of vehicles speed and acceleration.

Since it is possible to assess speed changes from audio data, we also follow this approach in the work presented in this paper. Additionally, audio data can be acquired in low visibility conditions, and such non-intrusive measurements allow observations of true habits of the drivers.

1.1 Proposed Approach

Although audio data are sometimes used in observations of vehicular traffic, still, to the best of our knowledge, no other researchers investigated extracting information on speed changes from audio data [25,26,29]. Most often, single speed measurements are performed, and they do not provide information about the dynamics of driving.

Audio data can be represented as spectrograms, showing frequency contents as a function of time, and then analyzed. Spectrograms may contain lines or curves corresponding to frequency components changing in time, as well as noises of various origin. In this paper, we work on extracting line parameters from spectrograms for audio data, recorded for cars passing by an audio recorder. These lines represent frequencies usually related to the car speed. Our goal is to detect whether the car is decelerating, accelerating, or maintaining stable speed.

Obtaining the information on speed changes from the spectrogram corresponding to audio data has numerous pitfalls. The lines in the spectrogram may be broken, they bent when the car is passing the microphone because of the Doppler effect, which is strongest when very close to the microphone, and they are on the noisy background. Our goal was to elaborate a methodology of extracting a simple representation of parameters of these lines (possibly broken, curvy and in noise). Image processing techniques were applied first to the spectrogram data. Next, we managed to use this very simple representation (i.e. the parameters of lines) of a complex spectrogram and bring classification rules that estimate the dynamic behavior of drivers.

1.2   Audio Data

The audio data we used were recorded on-road in controlled conditions, i.e. on an unfrequented road, to assure that the sound of the recorded car is not accompanied by sounds of other cars. This is because in this research we wanted to focus on the data representing a single vehicle. Mc Crypt DR3 Linear PCM Recorder, with 2 built-in microphones, was used to record stereo audio data, 48 kHz/24 bit. The audio data were recorded in the summer (August 2nd, 2016), winter (January 16th, 2017), and spring (March 31st and April 5th, 2017). Each data item represents a single drive, 10 s long, with the moment of passing the microphone in the center of the recording. Each drive represents one of 3 classes:

- acceleration, 111 drives,
- deceleration, 113 drives, and
- stable speed (with possible small, unintended variations), 94 drives.

A 300 m road segment was used for each drive. Speed changes were performed from about 60 m before to 60 m after passing the audio recorder. The description of the recordings is given below. Further details, including illustrations and the information on how to get the data, can be found in [29].

Summer Recordings. Summer recordings were made in Ciecierzyn, Lublin voivodship, in Poland, on a sunny day (weekday; maximum temperature this day was 24 °C), from 10 a.m. to noon. The road was in a broad mild basin, so the cars were not driving uphill nor downhill. The audio recorder was placed 1.5 m above the surface and as close to the road as possible. The recorder position was about 51°18'34"N, 22°36'13"E (GPS coordinates). The segment of the road used started at 51°18'39"N, 22°35'58"E, and ended at 51°18'30"N, 22°36'25"E. Three cars were recorded: 2 with Diesel engine (Toyota Corolla Verso and Skoda Octavia), and 1 with gasoline engine (Renault Espace). For each car, two drives per class were recorded, with additional 2 drives of Skoda. The audio data represent acceleration 50–70 km/h, deceleration 70–50 km/h, and stable speed of 50 km/h, plus 2 drives at 70 km/h for Skoda (20 drives altogether).

Winter Recordings. Winter recordings were made in the outskirts of a small town, Lubartów, Lublin voivodship, in Poland, from 6 p.m. to 8.30 p.m. [26]. The GPS coordinates of the recorder were 51°26'29"N, 22°35'59"E. The segment of the road used started at 51°26'35"N, 22°36'31"E, and ended at 51°26'26"N, 22°35'33"E. There was snow on the road, but not on the area below the tires. The temperature outside was −3 °C. One car was recorded: Renault Espace IV (2007), with manual transmission. The data represent 84 drives: 28 for acceleration 50–70 km/h, 28 for stable speed, 50 km/h, and 28 for deceleration 70–50 km/h, all without changing gear and without applying brakes (engine braking only).

We assured that the drivers did not change gear when passing by the microphone, as changing gear changes the lines in the spectrogram.

Spring Recordings. Spring recordings were also made at the same road near Lubartów, in 2 days in early spring, from 8.30 a.m. to 11.30 a.m. The GPS coordinates of the recorder and the road were the same as in the winter. The weather was windy on March 31st, and good on April 5th, with maximum daytime temperature 16 °C on March 31st and 19 °C on April 5th. The wind gusts did not affect the audio data, but to avoid strong wind gusts, a windscreen was applied later. The recorded cars included Renault Espace III and Espace IV, both with a gasoline engine, Skoda Octavia with a Diesel engine, and Smart ForFour with a gasoline engine, all with a manual transmission. The data include 214 drives, namely 77 for acceleration (50–70 km/h, and 50–80 km/h for Skoda), 58 for stable speed, at 60 km/h, 70 km/h, and 80 km/h, and 79 for deceleration (80–40 km/h, 80–50 km/h, and 70–40 km/h); brakes were applied here.

2   Methodology

Audio signal can be useful as a source of information about traffic, as it can be recorded at low visibility conditions, but it requires processing to extract this information. In our approach, we use spectrogram, i.e. the graph representing the frequency contents of sound as a function of time, as a basis of a graphic-based approach. Spectrograms are based on FFT (Fast Fourier Transform) spectrum, calculated for 170 ms frame, with 57 ms hop size (i.e. with 2/3 overlap), Hamming-windowed. In the preprocessing step the signal was low-pass filtered, and spectra for frequencies up to 300 Hz were used for preparing spectrograms. To facilitate further work, the audio data for each drive were represented as 4 spectrograms: 5 s before passing the microphone for the left channel, 5 s for the right channel, 5 s after passing the microphone for the left channel, and 5 s for the right channel. The choice of the length of the analyzed segments was arbitrary; we decided to analyze 5-s parts of the spectrogram, to ensure that the audio segments are long enough to capture speed changes. The linear frequency scale was used in spectrogram.

Spectrograms for automotive data, representing cars passing the road near the microphone, contain lines at low frequencies. These lines are mostly horizontal if the driver maintains approximately stable speed, rising up if the driver is accelerating, and descending if the driver is decelerating. Exemplary spectrogram is shown in Fig. 1; more spectrograms are shown in the left set of images in Fig. 2. We selected 5 s long one-channel segments with clearly visible lines. Our motivation was to have segments long enough to observe lines corresponding to changing speed (or maintaining constant speed), as for longer lines we can achieve better precision of calculating the parameters of the investigated lines.

When acquiring the audio data we did our best to assure that every time a single vehicle drive was recorded, representing a single target class, and it was not accompanied by sounds representing other vehicles. If the investigated

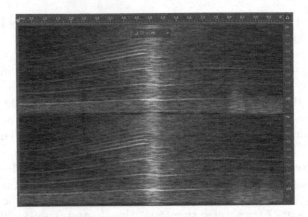

Fig. 1. Exemplary spectrogram (in grayscale), for a car recorded while accelerating. The upper and lower graphs represent the left and right channel. The center part represents the moment of passing the recorder

sounds were accompanied by other sounds, especially by the sounds of other vehicles passing by, then these accompanying sounds would be also represented as lines in the spectrogram. Therefore, if the other vehicle's sound was louder than the target sound, it could be indicated as an outcome of our algorithm.

In this research we focus on the strongest line in the spectrogram in the investigated spectrograms, but it can be extended to a multiple-vehicle case, via finding another strong lines in the spectrogram. The observed lines correspond to partials (harmonic or not) of pitched sounds, whose frequencies are usually related to the car speed. These frequencies change in time, and this is illustrated in the spectrogram. However, the spectrogram data are much more complex, as they also contain noises. Additionally, the lines are actually curves, especially at the moment of passing by the microphone. This is caused by the Doppler effect, most pronounced near the microphone (see Fig. 1).

In the presented approach, we aim to extract line parameters from these spectrograms. Our goal is to detect whether the vehicle is accelerating, decelerating, or maintaining approximately stable speed. Although the lines in the spectrogram may be broken, they bend when the car is passing the microphone, and there is a lot of noise in the spectrogram, we still believe that it is possible to extract parameters of these lines, and then this small set of parameters can represent a very complex spectrogram as indicator of speed changes.

The extraction of lines from the spectrogram can be based on edge detection algorithms, as the image analysis based on edge detection is insensitive to change in overall illumination level [30]. Edge detection in the image can be performed using e.g. Sobel operator. The lines of interest in our spectrograms are either horizontal, or slightly ascending or descending. However, as we can see in the last column of Fig. 2, the edges extracted by the Sobel operator do not represent our lines. Therefore, more sophisticated approaches must be elaborated.

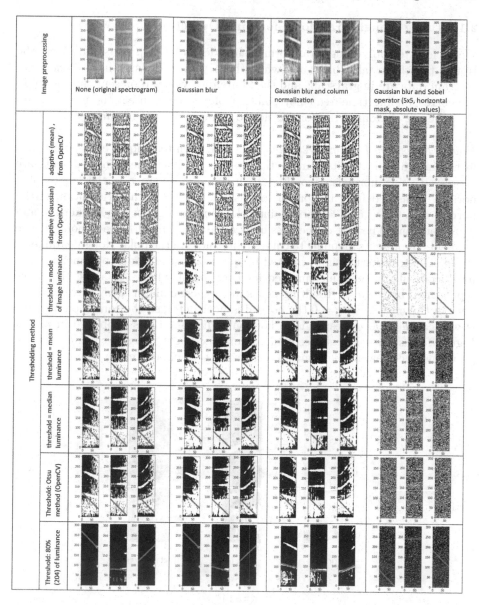

Fig. 2. Hough transform for spectrograms in grayscale, for various image processing methods and thresholds. The images in the cells of this table represent spectrograms for deceleration (left), stable speed (center), and acceleration (right). Gray lines on black-and-white images represent the results obtained from the Hough transform, i.e. θ and r corresponding to the strongest line, as indicated by the maximum of the accumulator array

2.1   Using Hough Transform for Line Detection

We decided to apply Hough transform for line detection [31]. In its main form, the Hough transform takes black-and-white (binary) images as an input. The spectrogram data are in grayscale, and the luminance values represent the energy in the corresponding time-frequency points. Color scales can also be used. Therefore, the use of Hough transform is not so straightforward in our case.

Hough Transform for Line Detection. In the Hough technique, each point (x,y) in the image indicates its contribution to the physical line. Line segments are expressed using normals: $x\cos(\theta) + y\sin(\theta) = r$, where r is the length of a normal, measured from the origin to this line, and θ is the orientation of the normal wrt. the x axis. For any point belonging to a given line segment, r and θ are constant. The plot of the possible r, θ values, defined by each point of line segments, represents mapping to curves (sinusoids) in the polar Hough parameter space. The transform is implemented by quantizing the Hough parameter space into accumulator cells, incremented for each point which lies along the curve represented by this r, θ. Resulting peaks in the accumulator array correspond to lines in the image. For $\theta = 0$ the normal is horizontal, so the corresponding line is vertical; $\theta = 90$ corresponds to horizontal line; and r is expressed in pixels. Figure 3 illustrates Hough accumulator matrix (right image), calculated for a grayscale spectrogram (left image), converted to a binary image (center image).

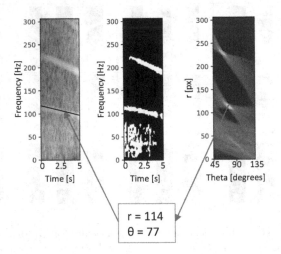

Fig. 3. The spectrogram in grayscale (left), in black-and-white (center), and the accumulator (right) for this spectrogram. The line marked in the left image corresponds to the maximum of the accumulator array

The Hough transform is applied on binary images. There are implementations that take grayscale images as an input, but then the image is transformed to binary. We applied various image processing techniques to obtain binary representations of the spectrograms, as shown in Fig. 2 and 4. We wanted to determine 2 approaches for detecting lines in the spectrogram using Hough transform:

1. threshold-based grayscale-to-binary conversion as input of the Hough transform - 1st approach, and
2. Canny edge detection [32] used as grayscale-to-binary conversion before applying the Hough transform - 2nd approach.

OpenCV implementation of the Canny algorithm was used [33]. In the preprocessing step, we tested Gaussian blur, column normalization, and Sobel operator in the 1st approach. Sobel operator was not used in the 2nd approach, as it is used as the edge detector operator in Canny algorithm. Gaussian blur was applied to get rid on noise, and column normalization was performed to obtain the same energy levels for each time point, as the energy at the moment of passing the microphone was much higher than in the remaining parts of the spectrogram. Column normalization consisted in rescaling each column of the spectrogram to the range $\{0, 255\}$, corresponding to 8-bit grayscale.

In our image processing (applied to the spectrograms), we tested several thresholding options, commonly used in similar works. Namely, 7 thresholding versions were used next for the 1st approach, i.e. in grayscale-to-binary conversion, and 3 for the 2nd approach, i.e. Canny edge detection, both as preprocessing before applying the Hough transform. These methods represent various approaches, commonly used in image processing [30].

The thresholds tested in the 1st approach included: 80% of the maximum luminance, Otsu method [34], median luminance of the spectrogram, mean luminance, mode of the luminance histogram, adaptive (Gaussian) threshold from OpenCV, and adaptive (mean) threshold from OpenCV. Thresholds tested in the Canny algorithm included the following threshold pairs:

– 20% and 80% of luminance; this threshold pair was based on [35],
– 0.66 of the mean luminance and 1.33 of the mean luminance,
– 0.66 of the median luminance value and 1.33 of the median luminance value.

Our goal was to find the preprocessing and thresholding that work best.

Two out of 7 thresholding versions in the 1st approach were based on options available in OpenCV, i.e. adaptive Gaussian threshold (based on Gaussian window), and adaptive mean threshold. The adaptive thresholds change locally, depending on the local luminance level. We also decided to choose other thresholding versions, adjusted to the mode of the image histogram, mean luminance, or median luminance; these thresholds are adjusted to the luminance of the whole image. These thresholds represent uniform thresholding, with one threshold applied to the converted image. The Otsu method, an optimal method described in [34], is also commonly used, so we decided to test it. However, the best result was obtained for a fixed threshold. We tested several version, namely from 10% to 90% of the luminance level, with 10% point step, and the 80% produced the best results. Testing was performed on several spectrograms;

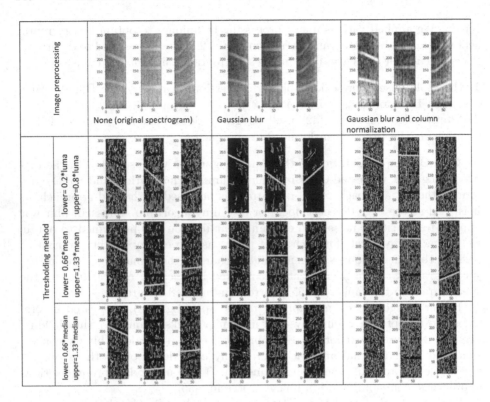

Fig. 4. Hough transform for various spectrograms versions, with Canny edge detector applied for grayscale-to-binary conversion, with various thresholds.

we choose the most difficult ones (i.e. with hardly visible lines). We first visually assessed the quality of the obtained black-and-white results, and next we also applied the Hough transform to the resulting images, in order to check (visually) if the indicated lines actually represent the target lines. After comparing all these results, we concluded that the fixed threshold of 80% of the luminance level works best.

In the 2nd approach, based on Canny edge detection, two thresholds are needed. The pixel is accepted as an edge, if its gradient is higher than the upper threshold, and the pixel is rejected if its gradient is below the lower threshold. The pixels between these thresholds undergo edge tracking by hysteresis thresholding. In the hysteresis thresholding step, pixels are added to edges if and only if at least one of the pixels around the one being processed represents an edge.

We tested 3 options of selecting the threshold pairs, again using OpenCV and options available in this implementation.

The analysis of the results of this processing, shown in Fig. 2 and 4, indicated which methods can be applied in the classification of these data into 3 classes, i.e. acceleration, deceleration, and stable speed. Next, we extracted parameters to represent lines in classification. As a result, we propose the following details of the 2 approaches of extracting spectrogram representations:

- Approach 1: Gaussian blur of the spectrogram, column normalization, and next threshold 80% of luminance in grayscale-to-binary image conversion are applied. Afterwards, the Hough transform is applied to find white lines in the processed image, for θ between 45 and 135°. A 2D array (accumulator) is calculated, and its maximum indicates θ corresponding to the strongest line in the spectrogram. For each of 4 spectrogram parts calculated for a single drive, the maximum of the accumulator and its corresponding θ and r constitute our set of parameters, i.e. 12 parameters represent one drive.
- Approach 2: Canny edge detection algorithm is used (on Gaussian blurred and column normalized spectrogram) as grayscale-to-binary image conversion method, instead of simple thresholding. To avoid tracking very low frequencies, for which lines are almost horizontal, we decided to limit the analyzed spectrogram to frequencies above 10 Hz. After a visual inspection of the obtained results of the Hough transform for the 3 tested Canny threshold sets, 0.66 and 1.33 of the median luminance were selected. Again, the maximum of the accumulator and its corresponding θ and r for each of 4 spectrogram parts constitute our 12-element feature set.

This way, a very simple representation of complex spectrograms can be used, namely the maximum of the accumulator and its corresponding θ and r for each 5 s segment of the spectrogram, for each channel of audio data (so, we have 12 features altogether for each drive). Next, we applied decision trees and random forests as classifiers for these data.

2.2 Heuristic Methodologies

Since we have a simple representation, we actually do not need complicated classification algorithms. Even more, we propose and then test 3 simple heuristic methodologies of classifying the underlying audio data into acceleration, deceleration, and stable speed classes, based on the 12-element feature vector:

1. We take θ corresponding to maximum accumulator of the 4 spectrogram parts for this sound. If $\theta > AccSlope$, the data are classified as acceleration, if $\theta < DecSlope$, then the data are classified as deceleration, and other values indicate stable speed. The thresholds $AccSlope$ and $DecSlope$ were experimentally chosen. Namely, we selected the thresholds based on testing θ values within $[45°, 135°]$, with 1° step, on the entire dataset. For Approach 1, $DecSlope$, and $AccSlope$ were equal to 80° and 90°, and for Approach 2 these slopes were equal to 87° and 95° respectively.
2. We take θ corresponding to the greatest r in the feature vector, and apply the same classification rule as in methodology 1. Again the boundary slopes were experimentally selected. For Approach 1, 81° as $DecSlope$ and 88° as $AccSlope$ were selected, whilst for Approach 2 these slopes were equal to

85° and 94°. Remark: It may look surprising that the θ below 90° indicates acceleration, but this can be caused by the Doppler effect (see Fig. 1), which decreases the frequency, and the lines are bent down. Note that the slope boundary values for methodology 1 may differ from the values for methodology 2, because in the latter the lines found for higher frequencies are preferred (r is greater in this case), and these lines are sloping more.

3. We also used θ and r values corresponding to the maximum of the accumulator to calculate a decision tree, in order to obtain an illustrative and well grounded classification rule. The conditions in the nodes of the tree indicate the boundary values at each step of this tree-based classification.

3   Experiments and Results

In our previous work [29] on the same audio data, we obtained up to 90.9% accuracy in ten-fold cross-validation (CV-10) using 85-element feature vector based on audio data, up to 90.3% when using 24-element feature vector describing lines in the spectrogram, and up to 92.6% when combining both feature vectors. No image processing was applied to the spectrogram data. Random forests, support vector machines, and multi-layer perceptrons were applied for data classification. Additionally, forest of shapelets were used, but the obtained accuracy was lower.

In this work, we propose using a very simple representation of a complex spectrogram, namely the parameters of the strongest line in each analyzed segment of the spectrogram (two parameters per segment, plus the maximum of the accumulator). Image processing techniques were applied to these spectrograms.

Figure 2 visually shows that the most effective procedure for using the Hough transform in Approach 1 is based on Gaussian blur and normalization for the threshold equal to 80% of the maximum luminance, so this method was chosen in further experiments with Approach 1. Figure 4 illustrates how using Canny edge detection method influences the results of Hough transform, when using the results of Canny method as the input of the Hough transform. As we can see, all tested threshold pairs gave the same results on exemplary blurred and normalized images, see Fig. 4. Thresholds based on the mean values may be better suited to the image content, and they seem to be more accurate than fixed threshold values. Mean is more affected by every single observation [36], including outliers. Therefore, the threshold pair based on median was applied in further experiments with Approach 2.

Decision tree, namely J4.8 tree from WEKA [37] was used as classification tool for the proposed spectrogram representation, i.e. using 12-element feature sets. Almost 80% accuracy was achieved for CV-10 cross-validation classification for Approach 1, and more than 73% for Approach 2. Confusion matrices for these cases are presented in Table 1. For the comparison with our previous work, random forest classifiers were also used. R package was applied for this purpose [38], yielding 85% accuracy in CV-10 for Approach 1 and almost 79% for Approach 2. As we can see, random forests yield better results than a single classification tree, especially for Approach 1. The results are 5–7% points lower than in our previous research, but here we have only 12 features, instead of up to 575 features. Table 2 shows the results obtained for both approaches through CV-10.

Table 1. Confusion matrices for the J48 decision tree classifiers and a) Hough line parameters from binary image obtained through 80% thresholding (Approach 1), b) Hough line parameters detected via Canny edge detection (Approach 2)

a) Classified as:	Dec	St	Acc	b) Classified as:	Dec	St	Acc
Dec	92	18	3	Dec	86	16	11
St	16	65	13	St	21	58	15
Acc	8	7	96	Acc	10	12	89

Table 2. Confusion matrices for the random forest classifiers and a) Hough line parameters from binary image obtained through 80% thresholding (Approach 1), b) Hough line parameters detected via Canny edge detection (Approach 2)

a) Classified as:	Dec	St	Acc	b) Classified as:	Dec	St	Acc
Dec	101	8	4	Dec	95	14	4
St	12	73	9	St	22	62	10
Acc	0	14	97	Acc	5	12	94

The decision tree constructed for the entire data set using Approach 1 classifies correctly 298 out of 318 objects (with 93.7% accuracy). When visually inspecting misclassified objects, we can observe that the lines are actually hardly visible in the spectrograms for these data.

Canny edge detection method used as preprocessing yielded worse results. We think that is because the Canny line detector is designed to detect lines on the background, whereas in the spectrogram there might be no background clearly separated from the target line. Additionally, the selection of thresholds may affect the quality of the representation.

After testing our approaches using decision trees and random forests, we also evaluated the proposed 3 simple heuristic methodologies of classifying the spectrograms based on the detected lines. The heuristic rules (1 and 2) take only one parameter (θ) into account. Cross-validation was not performed in this case. We are aware that the rules were extracted for the analyzed examples, but we believe that the rules are related to range of speed changes considered to represent stable or changing speed rather than to particular recordings.

The results of the heuristic methodologies no. 1 and 2, proposed in Sect. 2.2, are shown in Table 3, together with the results for Methodology 3 (CV-10 was applied in this case). The results for Methodology 1 are quite good, as just 80% was achieved using decision trees for 12-element feature vector using Approach 1.

Only 83% was achieved using the tree built for the entire data set, using Methodology 3 and Approach 1. The confusion matrices for Approach 1 are shown in Table 4 for Methodology 1 and Methodology 2, and in Table 5 for Methodology 3. As we can see, acceleration and deceleration are rarely confused, especially when using Methodology 1. The classification rules obtained via Methodology 3 in the form of the decision tree (built using the whole training set using Approach 1) are shown in Fig. 5. Note that the decision conditions in the top node and in the left subtree correspond to the formulas used in Methodology 1 for this approach.

Table 3. Obtained results. The highest accuracy is shown in bold

	Approach 1	Approach 2
Methodology 1	**79%**	79%
Methodology 2	71%	66%
Methodology 3	75%	76%

Table 4. Confusion matrices for the heuristic methodologies (a) 1 and (b) 2. Hough transform was performed on binary image obtained through uniform thresholding

a)
Classified as:	Dec	St	Acc
Dec	92	14	7
St	15	59	20
Acc	3	8	100

b)
Classified as:	Dec	St	Acc
Dec	97	10	6
St	21	58	15
Acc	24	16	71

Table 5. Confusion matrix for Methodology 3, obtained via CV-10 [37]

Classified as:	Dec	St	Acc
Dec	86	18	9
St	17	65	12
Acc	4	20	87

We can conclude that our proposed simple heuristic methodologies, based on just one parameter, offer results comparable with such classifiers as decision trees or random forests.

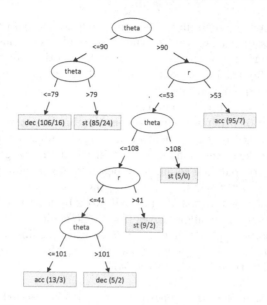

Fig. 5. The decision tree for the 3rd methodology

4   Summary

In this paper we aimed at elaborating a methodology of extracting a simple representation of automotive spectrograms. Our goal was to extract parameters of lines representing accelerating (in 40–80 km/h range, to discover speed changes around radars), decelerating, or maintaining stable speed. These lines are curvy, often broken, and accompanied by noise. Still, we managed to extract line parameters, and obtain a very simple representation, which allows detecting the behavior of drivers when passing the microphone or other measurements point, e.g. the radar. We proposed and tested several methodologies of extracting and representing lines in classification of speed changes.

In the presented work, we tested the Hough transform as a line detection tool, with spectrogram images as input data. Compared to our previous work, where we applied various hand-crafted techniques to detect lines in spectrograms, we achieved simpler representation of lines in spectrograms, albeit with decreased accuracy. Still, we believe that adjusting the settings of the pre-processing steps before applying the Hough transform may improve the obtained results; a multitude of these settings to tune can be considered a drawback of this work. The virtue of the presented approach is a very simple representation of lines in the spectrogram, with almost 80% accuracy obtained when using a single feature, i.e. θ (although we have to calculate other features to get the appropriate θ value). The drawback of the presented approach is that the methods presented in this paper may yield incorrect results if the lines are not clearly present in the spectrogram, which happens in some cases in our data.

The recognition accuracy still needs improvement. We plan to inspect thoroughly the misclassified examples, as the misclassification may be caused by distant sounds, interfering with the target sound. Additionally, we are planning to analyze thoroughly the impact of threshold values in the Canny edge detector on the classification quality in our future works.

Also, to avoid parameterizing curves caused by the Doppler effect and minimize the influence of other sounds, we consider limiting the analyzed sound segments, namely discard the moment of passing the microphone (with strong Doppler effect) and keep the remaining part in which the target sound is loud enough to mask accompanying sound. It is possible that a shorter segment (e.g. 4 s, 2 s before and 2 s after passing, excluding the moment of passing) will work better. However, the exact duration which will work best is to be found in further work. This work focused on single drives (for single vehicles), but line detection algorithms can also be applied when multiple cars are passing at the same time. Therefore, the same methodology can be adopted to the recordings of multiple vehicles and thus multiple lines in future work.

Acknowledgments. This work was partially supported by research funds sponsored by the Ministry of Science and Higher Education in Poland.

References

1. Elvik, R., Vaa, T.: The Handbook of Road Safety Measures. Elsevier, Oxford (2004)
2. Król, M.: Road accidents in Poland in the years 2006–2015. World Sci. News **48**, 222–232 (2016)
3. Huvarinen, Y., Svatkova, E., Oleshchenko, E., Pushchina, S.: Road safety audit. Transp. Res. Proc. **20**, 236–241 (2017)
4. Talebpour, A., Mahmassani, H.S., Hamdar, S.H.: Modeling lane-changing behavior in a connected environment: a game theory approach. Transp. Res. Proc. **7**, 420–440 (2015)
5. Banovic, N., Buzali, T., Chevalier, F., Mankoff, J., Dey, A.K.: Modeling and understanding human routine behavior. In: 2016 CHI Conference on Human Factors in Computing Systems, pp. 248–260. ACM, Santa Clara (2016)
6. Bonsall, P., Liu, R., Young, W.: Modelling safety-related driving behaviour-impact of parameter values. Transp. Res. A Policy Pract. **39**(5), 425–444 (2005)
7. Meiring, G., Myburgh, H.: A review of intelligent driving style analysis systems and related artificial intelligence algorithms. Sensors **15**(12), 30653–30682 (2015)
8. Wang, W., Xi, J., Chong, A., Li, L.: Driving style classification using a semisupervised support vector machine. IEEE Trans. Hum.-Mach. Syst. **47**(5), 650–660 (2017)
9. Mehar, A., Chandra, S., Velmurugan, S.: Speed and acceleration characteristics of different types of vehicles on multi-lane highways. Eur. Transp. **55**(1), 1–12 (2013)
10. Brooks, R.M.: Acceleration characteristics of vehicles in rural Pennsylvania. Int. J. Res. Rev. Appl. Sci. **12**(3), 449–453 (2012)
11. Schroeder, B.J., Cunningham, C.M., Findley, D.J., Hummer, J.E., Foyle, R.S.: ITE Manual of transportation engineering studies. Institute of Transportation Engineers, Washington, D.C., US (2010)

12. Gupta, P.K., Sharma, I.: Study of traffic flow in an entire day at a congested intersection of Chandigarh. J. Civ. Eng. Environ. Technol. **2**(12), 70–73 (2015)
13. Gaca, S., Kiec, M.: Speed management for local and regional rural roads. Transp. Res. Proc. **14**, 4170–4179 (2016)
14. Lingani, G.M., Rawat, D.B., Garuba, M.: Smart traffic management system using deep learning for smart city applications. In: IEEE 9th CCWC Proceedings, pp. 0101–0106. IEEE, Las Vegas (2019)
15. Leduc, G.: Road traffic data: collection methods and applications. Working Papers on Energy, Transport and Climate Change, vol. 1, no. 55 (2008)
16. Gajda, J., Sroka, R., Stencel, M., Wajda, A., Zeglen, T.: A vehicle classification based on inductive loop detectors. In: IMTC 2001, pp. 460–464. IEEE (2001)
17. Capobianco, S., Facheris, L., Cuccoli, F., Marinai, S.: Vehicle classification based on convolutional networks applied to FMCW radar signals. In: Leuzzi, F., Ferilli, S. (eds.) TRAP 2017. AISC, vol. 728, pp. 115–128. Springer, Cham (2018). https://doi.org/10.1007/978-3-319-75608-0_9
18. Ishida, S., Liu, S., Mimura, K., Tagashira, S., Fukuda, A.: Design of acoustic vehicle count system using DTW. In: Proceedings of the ITS World Congress, Melbourne, Australia, pp. 1–10 (2016)
19. Luvizon, D.C., Nassu, B.T., Minetto, R.: A video-based system for vehicle speed measurement in urban roadways. IEEE Trans. Intell. Transp. Syst. **18**(6), 1393–1404 (2016)
20. Nemade, B.: Automatic traffic surveillance using video tracking. Proc. Comput. Sci. **79**, 402–409 (2016)
21. Balid, W., Tafish, H., Refai, H.H.: Intelligent vehicle counting and classification sensor for real-time traffic surveillance. IEEE Trans. Intell. Transp. Syst. **19**, 1784–1794 (2018)
22. Smadi, A., Baker, J., Birst, S.: Advantages of using innovative traffic data collection techniques. In: 9th International Conference on Applications of Advanced Technology in Transportation, Chicago, IL, US (2006)
23. Adnan, M.A., Sulaiman, N., Zainuddin, N.I., Besar, T.B.H.T.: Vehicle speed measurement technique using various speed detection instrumentation. In: IEEE Business Engineering and Industrial Applications Colloquium, pp. 668–672. IEEE, Malaysia (2013)
24. Middleton, D., Gopalakrishna, D., Raman, M.: Advances in traffic data collection and management. Texas Transportation Institute Cambridge Systematics Inc., Washington, DC, USA (2002)
25. Kubera, E., Wieczorkowska, A., Słowik, T., Kuranc, A., Skrzypiec, K.: Audio-based speed change classification for vehicles. In: Appice, A., Ceci, M., Loglisci, C., Masciari, E., Raś, Z.W. (eds.) NFMCP 2016. LNCS (LNAI), vol. 10312, pp. 54–68. Springer, Cham (2017). https://doi.org/10.1007/978-3-319-61461-8_4
26. Wieczorkowska, A., Kubera, E., Koržinek, D., Słowik, T., Kuranc, A.: Time-frequency representations for speed change classification: a pilot study. In: Kryszkiewicz, M., Appice, A., Ślęzak, D., Rybinski, H., Skowron, A., Raś, Z.W. (eds.) ISMIS 2017. LNCS (LNAI), vol. 10352, pp. 404–413. Springer, Cham (2017). https://doi.org/10.1007/978-3-319-60438-1_40
27. Göksu, H.: Vehicle speed measurement by on-board acoustic signal processing. Meas. Control **51**(5–6), 138–149 (2018)
28. Kotus, J.: Determination of the vehicles speed using acoustic vector sensor. In: 2018 Signal Processing SPA, pp. 64–69. IEEE, Poznan (2018)
29. Kubera, E., Wieczorkowska, A., Kuranc, A., Słowik, T.: Discovering speed changes of vehicles from audio data. Sensors **19**(14), 3067 (2019)

30. Nixon, M.S., Aguado, A.S.: Feature Extraction & Image Processing for Computer Vision, 3rd edn. Academic Press Inc., Orlando (2012)
31. Fisher, R., Perkins, S., Walker, A., Wolfart, E.: Hypermedia Image Processing Reference. Wiley, West Sussex (2000)
32. Canny, J.: A computational approach to edge detection. IEEE Trans. Pattern Anal. Mach. Intell. **8**(6), 679–98 (1986)
33. OpenCV. https://opencv.org/. Accessed 14 June 2019
34. Otsu, N.: A threshold selection method from gray-level histograms. IEEE Trans. Syst. Man Cybern. **9**(1), 62–66 (1979)
35. RoboRealm. http://www.roborealm.com/help/RGB%20Filter.php. Accessed 26 Nov 2019
36. Chiang, C.L.: Statistical Methods of Analysis. World Scientific Publishing Co. Pte Ltd., Singapore (2003)
37. WEKA. https://www.cs.waikato.ac.nz/ml/weka/. Accessed 15 June 2019
38. The R Foundation. https://www.r-project.org/. Accessed 28 Nov 2019

Author Index

Printed in the United States
By Bookmasters